Writing Scientific Research Articles

Strategy and Steps

Margaret Cargill and Patrick O'Connor

Margaret Cargill BA, DipEd, MEd (TESOL)
Adjunct Senior Lecturer
School of Earth and Environmental Sciences
The University of Adelaide
South Australia 5005
Australia

Patrick O'Connor BSc, PhD
Visiting Research Fellow
School of Earth and Environmental Sciences
The University of Adelaide
South Australia 5005
Australia

WILEY-BLACKWELL

A John Wiley & Sons, Ltd., Publication

This edition first published 2009, © 2009 by Margaret Cargill and Patrick O'Connor

Blackwell Publishing was acquired by John Wiley & Sons in February 2007. Blackwell's publishing program has been merged with Wiley's global Scientific, Technical and Medical business to form Wiley-Blackwell.

Registered office
John Wiley & Sons Ltd, The Atrium, Southern Gate, Chichester, West Sussex, PO19 8SQ, UK

Editorial offices
9600 Garsington Road, Oxford, OX4 2DQ, UK
The Atrium, Southern Gate, Chichester, West Sussex, PO19 8SQ, UK
111 River Street, Hoboken, NJ 07030-5774, USA

For details of our global editorial offices, for customer services and for information about how to apply for permission to reuse the copyright material in this book please see our website at www.wiley.com/wiley-blackwell

Library of Congress Cataloguing-in-Publication Data

Cargill, Margaret.
   Writing scientific research articles : strategy and steps /
Margaret Cargill and Patrick O'Connor.
     p.   cm.
   Includes bibliographical references and index.
   ISBN 978-1-4051-8619-3 (pbk. : alk. paper) – ISBN 978-1-4051-9335-1
(hardcover : alk. paper) 1. Technical writing. 2. Research. 3. Science news. I. O'Connor, Patrick, 1967– II. Title.
   T11.C327 2009
   808′.0666–dc22

                                        2008042543

A catalogue record for this book is available from the British Library.

Set in 10.5/13pt Janson
by SPi Publisher Services, Pondicherry, India
Printed and bound in Singapore by Fabulous Printers Pte Ltd

05   2011

Writing Scientific Research Articles

Contents

Preface

Writing Scientific Research Articles is designed for early-career researchers in the sciences: those who are relatively new to the task of writing their research results as a manuscript for submission to an international refereed journal, and those who want to develop their skills for doing this more efficiently and successfully. All scientists are faced with pressure to publish their results in prestigious journals and all face challenges when trying to write and publish. This book takes a practical approach to developing scientists' skills in three key areas necessary for success:

- developing strategy: understanding what editors and referees want to publish, and why;
- developing story: understanding what makes a compelling research article in a particular discipline area; and
- using language: developing techniques to enhance clear and effective communication with readers in English.

The skills required for successful science writing are both science- and language-based, and skill integration is required for efficient outcomes. We are an author team of a scientist and a research communication teacher who have combined our perspectives and experience to produce an integrated, multidisciplinary approach to the task of article writing.

We have written the book both for those who write science in English as their first language and those for whom English is an additional language (EAL). Although a very high proportion of the research articles published worldwide currently appears in English, scientific research is an intensely international and intercultural activity in the twenty-first century, and authors come from a wide range of language and cultural backgrounds. This situation adds another layer to the challenges facing authors themselves, journal editors and referees, and those who teach and support EAL scientists. We hope the book will be relevant to all professionals involved with the practice of research article writing.

The book is designed for use either by individuals as a self-study guide, or by groups working with a teacher or facilitator. Readers can prepare their own manuscript step by step as they move through the book, or use the book as a preparation phase and return to relevant parts when the time comes to write their own paper and navigate the publishing process. Web support for the book is available at www.writeresearch.com.au, with additional examples and links to other resources.

The book has arisen out of fruitful collaborations at the University of Adelaide over many years, and especially out of our work with the Chinese Academy of Sciences since 2001. There are many people to thank for their contributions both to the approach and the book. First on the language end of the continuum must be Robert Weissberg and Suzanne Buker, whose 1990 book *Writing Up Research: Experimental Research Report Writing for Students of English* laid such an effective foundation in using the insights of the worldwide community of genre-analysis researchers as the basis of effective teaching about research article writing. Next are John Swales and his colleagues over the years, for their research output, their teaching texts, and their modeling of humble and rigorous curiosity as an effective way into the worlds of other disciplines. Then the team at Adelaide that has built from these bricks a context where the book could emerge: especially Kate Cadman, Ursula McGowan, and Karen Adams, and so many scientists over the years. For bringing the perspective and experience of scientists, particular thanks go to those who have taught with us in China: Andrew Smith, Brent Kaiser, Scott Field, Bill Bellotti, Anne McNeill, and Murray Unkovich. We also thank those who have supported the training programs where we have refined our practical teaching approach, particularly Yongguan Zhu and Jinghua Cao. And, of course, the many early-career authors, in Australia, Vietnam, Spain, and China, who have participated in our workshops and contributed their insights and enthusiasm to the development of the book.

Our warm thanks go also to the people who have helped with the production of the book itself: Sally Richards, Karen Adams, Marian May, and our editors at Wiley-Blackwell, Delia Sandford and Ward Cooper. Remaining errors and omissions must be down to us.

Margaret Cargill
Patrick O'Connor
September 2008

A framework for success

CHAPTER 1

How the book is organized, and why

1.1 Getting started with writing for international publication

Welcome to the process of writing your research results as a paper for submission to an international refereed journal! You may speak and write English as your first language, or as an additional language: we have written this book for all inexperienced authors of scientific papers, and for all authors wanting improved strategies for writing effective papers in an efficient way.

In this book we will use other terms as well as *paper* for what you are aiming to write: it may be called a *manuscript*, a *journal article*, or a *research article*. (See Chapter 2 for comments on other types of scientific article.) All of these terms are in use in books and websites providing information and advice about this type of document: this *genre*. The concept of genre is important for the way this book works, as we have based our approach in writing it on the findings of researchers who work in the field of genre analysis. These researchers study documents of a particular type to identify the features that make them recognizable as what they are.

One of the key concepts in use in this field of research is the idea of the *audience* for a document as a key factor in helping an author write effectively. Whenever you write any document, it is helpful to think first about your audience: whom do you see in your mind's eye as the reader of what you are writing? So we will begin now by thinking about the audience for a scientific research article.

Who is your audience?

Often the audience that you think of first is your scientific peers – people working in areas related to yours who will want to know about your results – and this is certainly a primary audience for a research article. However, there is another "audience" whose requirements must be met before your peers will even get a chance to see your article in print: the journal editor and referees (also called reviewers; see Chapters 3, 13, and 14 for more information). These people are often thought of as gate-keepers (or as a filter), because their role is to ensure that only articles that meet the journal's standards and requirements are allowed to

Writing Scientific Research Articles: Strategy and Steps, 1st edition. By M. Cargill and P. O'Connor. Published 2009 by Blackwell Publishing, ISBN 978-1-4051-8619-3 (pb) and 978-1-4051-9335-1 (hb)

enter or pass through. Therefore it can be useful from the beginning to find out and bear in mind as much information as you can about what these requirements are. In this book we refer to these requirements as referee criteria (see Chapters 3 and 14 for details), and we use them as a framework to help unpack the expectations that both audiences have of a research article written in English. We aim to unpack these expectations in two different but closely interrelated ways: in terms of

- the content of each article section and its presentation; and
- the English language features commonly used to present that content.

To do this, the book uses an interdisciplinary approach, combining insights from experienced science authors and referees about content, with those from specialist teachers of research communication in English about the language. Elements of language that are broadly relevant to most readers of the book will be discussed in each chapter. In addition, Chapter 17 focuses on ways in which users of English as an additional language (EAL) can develop the discipline-specific English needed to write effectively for international publication. This chapter can be studied at any stage in the process of working through the book, after you have completed Chapter 1.

1.2 Publishing in the international literature

If you are going to become involved in publishing in the international literature, there are a number of questions it is useful to consider at the outset: Why publish? Why is it difficult to publish? What does participation in the international scientific community require? What do you need to know to select your target journal? How can you get the most out of publishing? We consider these questions in turn below.

Why publish?

We have already suggested that researchers publish to share ideas and results with colleagues. These are some other reasons for publishing:

- to leave a record of research which can be added to by others;
- to receive due recognition for ideas and results; and
- to attract interest from others in the area of research.

However, there are two additional reasons that are very important for internationally oriented scientists:

- to receive expert feedback on results and ideas; and
- to legitimize the research; i.e. receive independent verification of methods and results.

These reasons underscore the importance of the refereeing process we discussed above. However, there are difficulties associated with getting work published: difficulties that operate for all scientists, plus some that are specific to scientists working in contexts where English is a foreign or second language, which together are known as EAL contexts.

In addition to the language-related barriers that spring to mind, it is also important to realize that writing is a skill, whatever the language. Many of the points covered in this book are equally important for EAL scientists and those who speak English as their first language.

Getting published is also a skill: not all writers are published. Some reasons for this fact include the following.

- Not all research is new or of sufficient scientific interest.
- Experiments do not always work: positive results are easier to publish.
- Scientific journals have specific requirements which can be difficult to meet: publishing is a buyer's market.

These issues will be addressed as you proceed through the book.

Another reason that researchers find the writing and publication process difficult is that communicating your work and ideas opens you up to potential criticism. The process of advancing concepts, ideas, and knowledge is adversarial and new results and ideas are often rigorously debated. Authors facing the blank page and a potentially critical audience can find the task of writing very daunting. This book offers frameworks for you to structure your thinking and writing for each section of a scientific article and for dealing with the publishing process. The frameworks provided will allow you to break down the large task of writing the whole manuscript into small tasks of writing sections and subsections, and to navigate the publishing process.

What does participation in the international scientific community require?

A helpful image is to think about submitting a manuscript to an international journal as a way of participating in the international scientific community. You are, in effect, joining an international conversation. To join this conversation, you need to know what has already been said by the other people conversing. In other words, you need to understand the "cutting edge" of your scientific discipline: what work is being done now by the important players in the field internationally. This means:

- getting access to the journals where people in the field are publishing;
- subscribing to the e-mail alert schemes offered by journal publishers on their websites so that you receive tables of contents when new issues are published; and
- developing skills for searching the Internet and electronic databases in libraries to which you have access.

Without this, it will be difficult to write about your work so as to show how it fits into the progress being made in your field. In fact, this knowledge is important when the research is being planned, well before the time when the paper is being written: you should try to plan your research so it fits into a developing conversation in your field.

Active involvement in international conferences is an important way to gain access to this international world of research in your field. Therefore you need both written and spoken English for communication with peers. This book aims to help with the written language, and some ideas for developing spoken science English are given in Chapter 16. As you become a member of the international research community in your field in these ways, you will develop the knowledge base you need to help you select the most appropriate journal for submission of your manuscript: we call this your *target journal*.

What do you need to know to select your target journal?

- Does the journal normally publish the kind of work you have done? Check several issues and search the journal website, if it has one. It is helpful if you can cite work from the journal in the Introduction of your manuscript, to show that you are joining a conversation already in progress in the journal.
- Does the journal referee the papers? This is absolutely imperative for enhancing the international credibility of your work. It may also be important to check the journal's impact factor, if this measure is important for assessing research outcomes in your country or research context. (See Chapter 12 for more information on impact factor, citation index, and other similar measurements.)
- Does the journal publish reasonably quickly? Many journals include the dates when a manuscript was received and published underneath the title information, so you can check the likely timeline. Others include this information on their websites.
- Are there page charges? Some journals charge authors a fee to publish, or to publish coloured illustrations. Check whether this is the case. If so, you can ask whether the journal is willing to waive these charges for authors in some parts of the world.
- Are members of the editorial staff efficient and helpful? Some journals have information on their website with targeted advice for authors from EAL backgrounds, or you may be able to ask colleagues who have submitted to particular journals about their experiences. It can be especially useful to share this kind of information among colleagues in your laboratory group or work team, perhaps as part of a program to encourage international publication of the work of your institution or group.

More detail about evaluating different journals and selecting your target journal is given in Chapter 12.

How can you get the most out of publishing?

Publishing quickly is often helpful. In addition, publishing in a widely read journal is better for you (higher citation index; see Chapter 12). However, if you aim too high in relation to the international value of the work you have done, you may be rejected, and resubmission takes more time. These two issues have to be balanced carefully to determine an optimal strategy for your own situation. Finally, publishing where your peers will read the paper is important.

Once you have thought about the issues raised above, and made some preliminary decisions about a possible target journal, you are ready to move on to consider the aims of this book.

1.3 Aims of this book

The aims of the book are to provide you, the reader, with:

- an improved understanding of the structure and underlying logic of scientific research articles published in English in the international literature;

- an overall strategy for turning a set of results into a paper for publication;
- skills for analysing the structure and language features of scientific articles in your own discipline, and for using the results of this analysis to improve your own scientific writing;
- knowledge of the stages involved in the process of submitting an article for publication, and strategies for completing each stage;
- knowledge and basic mastery of the specific English language features commonly used in each section of published articles;
- strategies and tools for improving your own drafts, such as structured checklists, ways to strategically re-use relevant language elements, special-purpose software, and discipline-specific writing groups; and
- a process for completing a draft of an article on your own research results, prepared in the style of the journal to which you wish to submit.

1.4 How the book is structured

Two principles underlie the way we have organized this book: that people learn best by doing, and that you will want to continue developing your skills on your own or with colleagues in the future, even if you first encounter the book in a classroom environment. Therefore we aim to show you how you can use examples of journal articles, from your own field and also from others, to learn more about writing for publication.

To achieve this goal, the book will often invite you to discuss examples with a colleague and then report to a larger group. This assumes that you are using the book in a class situation. However, if you are using it for individual study, you can note down your answers and then revise them once you reach the end of a section. As we move through the book, you will also have the opportunity to draft (or substantially revise) your own article, section by section, if this is appropriate.

Instructions for activities in the book will use the following terms to refer to different categories of example articles:

- Provided Example Article(s) (PEAs): these are two articles chosen by the authors of the book and included in full at the back (Chapters 18 and 19). You will use both in the early sections of the book and then be asked to select one to use in more detail.
- Selected Article (SA): this is an article that you will choose from your own field of research, and that may be from your target journal. You will choose your SA as you continue with Chapter 1.
- Own Article (OA): this is the draft manuscript you will write using your own results as you progress through the book. If you do not yet have your own results, you can skip the tasks relating to the OA and come back to them later.

The following sections of the book work like this.

- We present information about the structure of research articles, section by section, which has been summarized from the work of scholars in the field of applied linguistics over the last 20 years. We present this as a *de*scription, not a *pre*scription: i.e. "this is what the scholars have found", not "this is what you should do". We do this because there are many effective ways to write articles,

---

Task 1.1 Selecting an article to analyze

Select an article in your own field of research to use as your SA (Selected Article), preferably from your target journal and preferably written by a native speaker of English (check authors' names and the location of their work sites to help identify an author's language background). We suggest that you do not choose your SA from *Nature* (UK) or *Science* (USA), as these two journals use conventions that are very different from most other journals. It will be more useful to learn the more usual conventions first, and then adapt them later if you need to. (See Chapter 2 for more details on the differences in article structure.)

---

not just one way. Our aim is to help you develop a repertoire (a range of effective possibilities) to select from, depending on the goals you have for a given article section.

- Then we ask you to look at the relevant section of the PEA (Provided Example Article) and check whether you can find the described features there (answers to the Tasks can be found in the Answer pages at the end of the book).
- Next, we ask you to analyse your own SA for the same features, and think about possible reasons for what you find.
- Finally, we ask you to work on the draft of your OA (Own Article), using the new information you have gained from the analysis. (These sections are optional for readers who do not have their own results ready to write up.)
- As well as this analysis of structural features, the book includes teaching, analysis, and exercises on elements of English language usage that are particularly relevant to each section of a research article. Again, answers are in the Answer pages. If English is your first language, you may choose to skip some or all of these sections.
- After all the sections of a research article have been covered in this way, we focus on the process of submitting the manuscript to the journal, and how to engage in correspondence with the editor about possible revisions.
- Chapter 15 summarizes a process for preparing a manuscript from first to last, with strategies for editing and checking.
- Chapter 16 focuses on techniques and strategies for ongoing development of your skills for writing, publishing, and presenting your research in English.
- Chapter 17 provides advice about specific features of science writing that often cause problems for authors with EAL. It can be studied at any stage of a reader's progress through the book.
- The final section of the book (Chapters 18 and 19) contains the two PEAs. Additional examples may be found on our website at www.writeresearch.com.au.
- At the end of the book you will find answers to the tasks that appear in the other chapters, and the Reference list.

CHAPTER 2

Research article structures

We will now look at the overall structure of research articles in science. In general, this follows a set of conventions that have developed over the years from 1665, when the first issue of *Philosophical Transactions* appeared in England. It is important to recognize that, within a common core structure, there are variations from field to field and from journal to journal: always check the specific requirements of your target journal before finalizing the structure of any article you write.

Before we look at the results of research into article structure, complete the introductory task below.

Task 2.1 Article headings and subheadings

Read quickly to find the headings of the sections of the PEAs (Chapters 18 and 19):

- How is each paper organized?
- What are the main headings and subheadings? Make brief notes.

Check your answers in the Answer pages.

Now look at the headings of your SA (a Selected Article from your own research field) and the SA of a colleague. Note the similarities and differences you find.

2.1 Conventional article structure: AIMRaD (Abstract, Introduction, Materials and methods, Results, and Discussion) and its variations

Before we explore article structure in detail, it is important to note that our focus in this book is on research articles based on experimental research. Other research paradigms, for example in humanities and social science fields, use different structures for their papers. Similarly, papers other than research articles use different structures. Of particular relevance to scientists are review articles (or reviews), which do not present new data from fresh experimentation, but rather selectively discuss and compare the findings of other scientists, in order to advance thinking in the area of interest. We will think more about these other types of scientific article in later subsections. First, we will consider the hourglass

Writing Scientific Research Articles: Strategy and Steps, 1st edition. By M. Cargill and P. O'Connor. Published 2009 by Blackwell Publishing, ISBN 978-1-4051-8619-3 (pb) and 978-1-4051-9335-1 (hb)

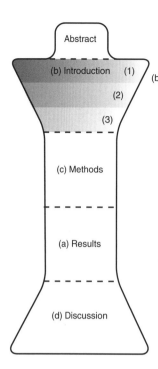

(a) The whole structure is governed by the Results box; everything in the article must relate to and be connected with the data and analysis presented in the Results section.

(b) (1) The Introduction begins with a broad focus. The starting point you select for your Introduction should be one that attracts the lively interest of the audience you are aiming to address: the international readers of your target journal.

(3) The Introduction ends with a focus exactly parallel to that of the Results; often this is a statement of the aim or purpose of the work presented in the paper, or its principal findings or activity.

(2) Between these two points, background information and previous work are woven together to logically connect the relevant problem with the approach taken in the work to be presented to address the problem.

(c) The Methods section, or its equivalent, establishes credibility for the Results by showing how they were obtained.

(d) The Discussion begins with the same breadth of focus as the Results – but it ends at the same breadth as the starting point of the Introduction. By the end, the paper is addressing the broader issues that you raised at the start, to show how your work is important in the 'bigger picture.'

Fig. 2.1 AIMRaD: the hourglass "shape" of a generic scientific research article and key features highlighted by this shape.

diagram (Figure 2.1) commonly used to represent the structure of an AIMRaD article, and what it can tell us about English-language research articles. In this diagram, it is the width and shape of the segments, rather than their depth, that tell us something important about scientific articles.

Here we represent an experimental article in terms of different component shapes put together into an hourglass configuration. This enables us to highlight several important features of such articles in a way that is easy to remember. The right-hand part of Figure 2.1 summarizes the features to focus on at this stage.

Task 2.2 Does the diagram match your understanding?

Discuss: Does this hourglass shape also represent the understanding of a research article in your culture or workplace? If not, can you suggest a diagram that shows how your understanding of a research article is different?

Of course, not all scientific research articles follow the simple structure given in Figure 2.1. There are two major variations that we will introduce here; these are presented visually in Figures 2.2 and 2.3. Study these figures now, before doing Task 2.3.

Other research article formats

The highly cited journals *Nature* (UK) and *Science* (USA) use variations of the common conventions for their article categories, reflecting the fact that their aim

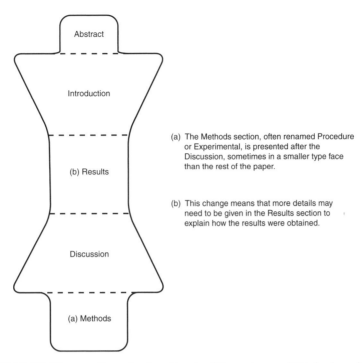

Fig. 2.2 AIRDaM (Abstract, Introduction, Results, Discussion, and Methods and materials): a structure variation that occurs in articles in some journals with a focus on molecular biology.

(a) The Methods section, often renamed Procedure or Experimental, is presented after the Discussion, sometimes in a smaller type face than the rest of the paper.

(b) This change means that more details may need to be given in the Results section to explain how the results were obtained.

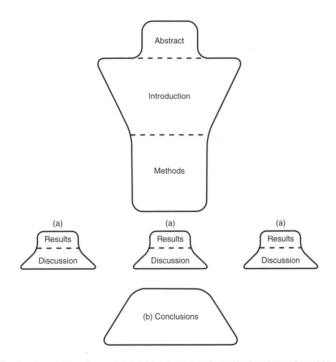

(a) The Results and Discussion are presented together in a single combined section; each result is presented, followed immediately by the relevant discussion.

(b) This change means that a separate section is needed at the end to bring the different pieces of discussion together; it is often headed Conclusions.

Fig. 2.3 AIM(RaD)C (Abstract, Introduction, Materials and methods, repeated Results and Discussion, Conclusions): a structure variation that is permitted in some journals, usually for shorter articles.

Task 2.3 Structure of the PEAs

Check the notes you made in answer to Task 2.1.

- Which of the three structures presented so far matches most closely the structure of the PEAs? (Check your answer in the Answer pages.)
- Which most closely matches your SA?

---

is to present highly significant new advances in science in ways that are very accessible to scientists who are not necessarily specialists in the areas covered by the articles. These articles typically begin with a carefully structured initial section introducing the background and rationale of the work to the wide range of expected readers, followed by a concise report of the findings and a short discussion. Methods are often only summarized in the main article, with full details appearing on a linked website. Full details on the structures required by these journals can be found on the journals' websites. Competition for publication in these journals is intense, and they are not likely to be realistic targets for most beginning scientists. For this reason we do not focus on their structure in this book.

Many journals offer alternatives to the article format for reporting research findings. Important among these are brief notes (also called research notes or notes), and letters. These may not include any section headings at all, but if you read them with an analytical eye you will be able to find the same types of information as are contained under the conventional AIMRaD headings in a full article.

Task 2.4 Prediction

Identify which part of a research paper the following phrases came from. Write one of the following letters at the end of each line: I = Introduction, M = Materials and methods, R = Results, or D = Discussion.

Example: It is very likely that... because...   (D)

...yielded a total of...   ( )
The aim of the work described...   ( )
...was used to calculate...   ( )
There have been few long-term studies of...   ( )
The vertical distribution of...was determined by...   ( )
This may be explained by...   ( )
Analysis was carried out using...   ( )
...was highly correlated with...   ( )

Check your answers in the Answer pages.

Now we begin to think in more detail about what information appears in the different sections of a research article. It is likely that you already know quite a lot about this, from reading articles for your own work. Task 2.4 focuses on this pre-existing knowledge.

It is likely that the clues you used to help you answer the questions in Task 2.4 related both to the vocabulary in the phrases and to elements of the grammar, especially the tense of the verbs (simple past, present perfect). We will build on this knowledge in later sections.

In Chapter 3 we will consider the relationship between the structure of research articles and the expectations of the gatekeeper audience that you, as an article submitter, are aiming to meet. The conventional structures we have been looking at in Section 2 have been maintained in science journals for a long time: we can assume that they must still serve the purposes of the journal editors effectively, and meet the needs of the journal readers. It is interesting to think about how and why that is the case.

CHAPTER 3

Referees' criteria for evaluating manuscripts

As discussed in Chapter 1, the first audience for your manuscript is the editor of the journal you have selected. In recent years, with the advent of electronic submission by uploading files on a computer, the very first audience may be a person who checks that formatting and other requirements have been met, but this fact does not alter the editor's initial filtering role in terms of the article's content. If the manuscript is judged suitable for refereeing (see Chapters 13 and 14 for more details of this process), the editor sends it to (usually) two peer reviewers or referees for comment. These referees are probably working in the same field as the manuscript authors: perhaps their names are in the list of references of the manuscript. However, the refereeing process is "blind", meaning that the manuscript authors do not know who reviews their paper. (Double-blind refereeing, where the referees also do not know who authored the manuscript they are reviewing, is less commonly practised in the sciences.)

Each journal has its own set of instructions for referees and sometimes these are available on the journal's website. You should check and see whether this is the case for the journal you are targeting, and obtain a copy if possible. For the purposes of this book, we have constructed a composite list of referee criteria that includes the sorts of questions referees are commonly asked to respond to (Figure 3.1). In addition to "ticking the boxes" to provide yes/no answers to the questions, referees are asked to write their comments about any problems with the manuscript or any suggestions for improvement that need to be followed before the manuscript can be considered suitable for publication in the journal. Increasingly, as the number of manuscripts submitted to journals has grown, referees are asked to give some numerical rating of the paper's novelty or quality as well (e.g. Does this manuscript fall within the top 20% of manuscripts you have read in the last 12 months?). Referees return their comments to the editor. Complete Task 3.1 now.

As we discuss each section of a research article in detail, we will keep these referee criteria in mind, and pay attention to the presentation features and English expressions that are commonly used to highlight the fact that evidence relevant to referee criteria is being presented.

We will begin by considering the question: Does the title clearly indicate the content of the paper?

Writing Scientific Research Articles: Strategy and Steps, 1st edition. By M. Cargill and P. O'Connor. Published 2009 by Blackwell Publishing, ISBN 978-1-4051-8619-3 (pb) and 978-1-4051-9335-1 (hb)

Typical questions included on Referee's Evaluation Forms for science journals

1. Is the contribution new?

2. Is the contribution significant?

3. Is it suitable for publication in the Journal?

4. Is the organization acceptable?

5. Do the methods and the treatment of results conform to acceptable scientific standards?

6. Are all conclusions firmly based in the data presented?

7. Is the length of the paper satisfactory?

8. Are all illustrations required?

9. Are all the figures and tables necessary?

10. Are figure legends and table titles adequate?

11. Do the title and Abstract clearly indicate the content of the paper?

12. Are the references up to date, complete, and the journal titles correctly abbreviated?

13. Is the paper excellent, good, or poor?

Fig. 3.1 Typical questions that referees are asked to answer when reviewing manuscripts for science journals.

Task 3.1 Where would referees look?

Read the list of questions in Figure 3.1. For each question, decide where in a manuscript a referee would expect to find evidence on which to base their answer. Write one or more of the following abbreviations beside each question: A, I, M, R, D, or Ref (meaning reference list). For example, for question 5 you would write *M and R*.

Check your answers in the Answer pages.

3.1 Titles as content sign posts

Good titles clearly identify the field of the research, indicate the "story" the results tell, and raise questions about the research in the mind of the reader. We will return to a more detailed consideration of titles in Chapter 10. For now, consider this example.

Title: Bird use of rice field strips of varying width in the Kanto Plain of central Japan

Information:
The focus is on birds in relation to rice fields.
The width of rice field strips was varied in the study.
Width of strips was correlated with the number and species of birds using them.
The research took place in central Japan.

Possible questions:
Why was the width of the strips an important variable?
Did the width of the rice field strips affect which birds used it?
If so, which field strip width was used most by which birds?
How did the researchers measure bird use?
Would the experiment be worth repeating for rice field strips in other places?

Task 3.2 Information extracted from titles

Look at the following titles and list the information about the research and its results you can deduce from the titles. What questions might you, as a reader, expect to answer by reading the article? (The questions will depend on the individual reader's reason for reading the text.)

Title A: Use of *in situ* ^{15}N-labelling to estimate the total below-ground nitrogen of pasture legumes in intact soil-plant systems

Information:

Questions:

Title B: Short- and long-term effects of disturbance and propagule pressure on a biological invasion

Information:

Questions:

Title C: The soybean NRAMP homologue, GmDMT1, is a symbiotic divalent metal transporter capable of ferrous iron transport

Information:

Questions:

Check your answers with the suggestions provided in the Answer pages.

Choosing one of the example articles as your focus for analysis tasks

Titles B and C above are the titles of the PEAs included at the back of the book. You will need to select one of them to use as the basis of text analysis exercises as we proceed through the sections of the book. The answers you gave to the questions in Task 3.2 should help you to decide which of these two articles will be more interesting and relevant to you.

Task 3.3 Unpacking the title of your SA

Now, repeat Task 3.2 for the title of your SA.

Title:

Information:

Questions:

When and how to write each article section

CHAPTER 4

Results as a "story": the key driver of an article

Because the results govern the content and structure of the whole article, it is important to be as clear as possible about the main points of your results "story" at the beginning of the writing process. We suggest that your first task when preparing to write a paper is to identify from your results a clearly connected story which leads to one or more take-home messages. This term refers to what readers remember after they have put the paper down: what they talk to their colleagues about over a cup of coffee next day, for example.

To move towards this clear story, focus on your tables and figures first. For each one, write a list of one or two bullet points highlighting the main message(s) of the data presented. Sort the figures and tables into the best order to connect the pieces of the story together. Draft some bullet points into a list to form a take-home message. Then sit down with all your co-authors and discuss the story of the paper that you will write. Aim to reach agreement on:

- which data should be included;
- what are the important points that form the story of the paper; and
- what is/are the take-home message or messages.

Task 4.1 Questions to focus the drafting process

Answer the four questions below, in English even if it is not your first language, for the results you want to turn into a paper.

1 What do my results *say*? (two sentences maximum, a very brief summary of the main points, no background!)
2 What do these results mean in their context? (i.e. what conclusions can be drawn from these results?)
3 Who needs to know about these results? (i.e. who specifically forms the audience for this paper you are going to write?)
4 Why do they need to know? (i.e. what contribution will the results make to ongoing work in the field? Or, what will other researchers be missing if they haven't read your paper?)

Writing Scientific Research Articles: Strategy and Steps, 1st edition. By M. Cargill and P. O'Connor. Published 2009 by Blackwell Publishing, ISBN 978-1-4051-8619-3 (pb) and 978-1-4051-9335-1 (hb)

Then you are ready to write the various sections of the manuscript itself.

We have found Task 4.1 useful in helping authors identify some key information that will help them begin the drafting process.

Once you can answer these questions for your own results, you are ready to refine your tables and figures so that they present, as clearly and forcefully as possible, the data that support the components of your story. That refinement process is the topic of Chapter 5.

Results: turning data into knowledge

The data presentation in a scientific article aims to illustrate the story, present evidence to support or reject a hypothesis, and record important data and meta-data. We verify, analyse, and display data to share, build, and legitimize new knowledge. To do this effectively we must present all necessary data in ways which make the most important points most prominent. Data presentation is also an exercise in deciding which datasets or details to leave out of the article. If you have decided to include figures or tables, they should be numbered and presented sequentially and referred to in that order in the text.

Many journals now accept additional data which support or extend the story as appendices or supplementary online data. For each data element in your paper you should ask yourself if it is necessary to the story of the paper, or not essential but valuable for those who might access it in an online archive. Remember, the referees will be asked to comment on whether all the tables and figures are necessary, and this will include the supplementary material.

Data presentation styles vary with discipline and personal preference and change over time, and there is a large amount of contradictory published advice about what to do, and what looks good. Our aim in this section is not to provide a concrete set of rules for data presentation but rather to help you optimize the presentation of your data to support the story of your article. One over-arching guideline is that tables and figures should "stand alone": that is, the reader should not need to consult the text of the article to understand the data presented in the table or figure; all necessary information should appear in the table/figure, in the title/legend, or in keys or footnotes.

The first reference for style of data presentation is the *Instructions to Contributors* (sometimes called Instructions to Authors or Author Guidelines, or other similar names) of the journal you intend to submit the article to. Not all Instructions to Contributors provide great detail about data presentation, but they will generally guide you in formatting and preferred style. The next best source of information on data presentation style is articles in recent issues of the journal. You can maximize your chances of meeting the journal's requirements by analysing the types of data presented, the choice of figures or tables, the choice of figure type, and the amount of data presented in the text and in the titles and legends. Use the results of your analyses to inform your decisions on the data presentation for your own manuscript.

Writing Scientific Research Articles: Strategy and Steps, 1st edition. By M. Cargill and P. O'Connor. Published 2009 by Blackwell Publishing, ISBN 978-1-4051-8619-3 (pb) and 978-1-4051-9335-1 (hb)

The choice of whether to use a figure, table, or text depends on the point or meaning you want the reader to receive from those data. Each form of data display has strengths and weaknesses.

Tables are most useful for

- recording data (raw or processed data);
- explaining calculations or showing components of calculated data;
- showing the actual data values and their precision; and
- allowing multiple comparisons between elements in many directions.

Figures are most useful for

- showing an overall trend or "picture";
- comprehension of the story through "shape" rather than the actual numbers; and
- allowing simple comparisons between only a few elements.

The choice is summarized in Table 5.1.

Table 5.1 The choice between data display in figures or tables.

Most useful	Table	Figure
When working with	number	shape
When concentrating on	individual data values	overall pattern
When accurate or precise actual values are	more important	less important

5.2 Designing figures

Design each figure around the point you want to get across most strongly. In an era when authors have access to many computer graphics packages and the ability to produce numerous graphical representations and styles, it is important to take charge of the software and direct it to your purpose. It may be helpful to determine the design elements you want in the figure before going to the graphics package. This will help you avoid using default settings or template styles which do not meet your needs. In designing your figures you may consider things such as

- which variable needs to have the most prominent symbol or line (heaviest line weighting);
- whether you want to emphasize differences or similarities between elements; and
- what scale, scale intervals, maximum and minimum values, and statistical representations are most meaningful.

The range of common figure types listed below allows you to emphasize different qualities of the data.

- **Pie charts** are effective at highlighting proportions of a total or whole.
- **Column and bar charts** are effective for comparing the values of different categories when they are independent of each other (e.g. apples and oranges).

- **Line charts** allow the display of a sequence of variables in time or space or the display of other dependent relationships (e.g. change over time).
- **Radar charts** are useful when categories are not directly comparable.

You should also be consistent with styles of figures throughout the article. It is especially important to keep the same symbols and order for given treatments or variables in all figures if possible. Also, keep figures free from clutter; too many different elements can distract the reader from the main points.

The journal may shrink your figure to fit the journal page or column width, and trendlines and symbols may become crowded and less distinct if they are not chosen carefully. Shrink your figures to the standard size for the journal you intend to submit your manuscript to, and check that all important features of your figure are still clear and obvious.

Figures are most appealing to the eye when they

- have 3:2 proportions;
- are boxed when there is relatively little ink in the figure; or
- are unboxed if there are numerous lines, bars or columns.

A review of figures in published articles shows a number of common weaknesses which reduce the power of figures to contribute to the communication of the story:

- the wrong figure type has been chosen and relationships between elements are not obvious when they are important or are apparent when they do not exist;
- weak descriptive titles are used when a story-telling title would be appropriate (many of the points discussed in Chapter 10 on article titles apply to titles for figures as well);
- data already shown in the text or tables are repeated in the figure;
- the shape, shading, pattern or weight of symbols, markers, or lines does not emphasize the main results or the story of the figure;
- the figure is unnecessarily cluttered with lines, legend symbols, numbers, or poorly chosen axis scale divisions;
- axes are not labelled descriptively or are labelled with the jargon of the scientific subdiscipline or research group;
- numbers are included when the exact values are not important to the story and the approximate values can be derived from the x and y axes; and
- data categories are not sorted to show priorities or important relationships between elements or the design of related figures is not consistent enough to allow rapid appraisal.

Small changes in the details of a figure can improve the communication of the main message. Figures 5.1 and 5.2 illustrate some improvements that can be made in a figure which already contains the necessary information but is not sharply focused on communicating the stand-alone message.

Improvements in Figure 5.2 in comparison with Figure 5.1 are listed below.

- Removal of error bars and replacement with LSD bar decreases clutter, allows comparison of significant differences between treatments and allows the y axis to be expanded with a lower maximum (i.e. greater spread between the lines). More detail about the significance level of difference is also provided in the figure legend. The removal of the figure border also reduces clutter in this line graph.

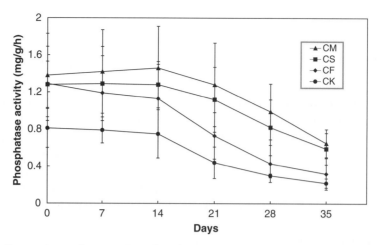

Fig. 5.1 Comparisons of root surface phosphatase activity of wheat plants for Control (CK), exclusively chemical fertilizer (CF), combined application of chemical fertilizer and wheat straw (CS), and farmyard manure (CM) treatments. Error bars represent the standard error of the mean for each treatment.

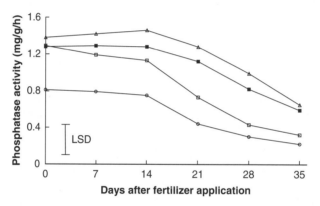

Fig. 5.2 Root surface phosphatase activity of wheat plants differed after soil amendment with different fertilizer treatments. Phosphatase activity was highest in farmyard manure (▵) treatments followed by combined application of chemical fertilizer and wheat straw (■), chemical fertilizer alone (□), and control/no amendment (○) treatments. Phosphatase activity declined over 5 weeks for all treatments. Least significant difference (LSD; two-way ANOVA, P ≤ 0.05) is 0.39 mg/g/h.

- The main comparison between chemical fertilizer and chemical fertilizer plus wheat straw is clearer as the same open and closed symbol is used (square) and other treatments can be compared with these two.
- Describing symbols in the figure legend instead of using an inserted legend leaves more white space to help readers compare the lines.
- The *x* axis is more descriptively titled and units are more appropriately spaced.
- The title has changed from a descriptive statement to a story-telling statement of what the data show.

There are other forms of figures which are not presentations of the results but demonstrate process (e.g. flow chart), methodology (e.g. apparatus), or documentary evidence, which may have been collected originally as a visual image (e.g. photograph or spatial representation). All of these forms should conform to the same basic rules as those discussed for figures above:

- highlight the most important information most prominently;
- be essential to telling and supporting the story with evidence; and
- be clear and consistent in style and do not duplicate data already presented.

Task 5.1 Examining data display

Examine your SA for the types of data and how they are displayed.

- Is the overall picture or trend obvious in the way the data are presented? Could it have been made more prominent?
- What comparisons between elements interest you, and does the presentation type and style make these comparisons easy?
- Are the necessary details of datasets presented to allow you to make calculations from the data?
- Does the figure have any of the weaknesses described above and how do these detract from the telling of the story?

5.3 Designing tables

Tables are often used to record data and meta-data of a study and may contain a number of rows or columns which require careful reading by the user before the meaning can be appreciated. This is especially true where tables contain a large number of cells and where comparisons between different rows and columns are necessary to understand the story. These potential limitations of tables can be largely overcome by good design, particularly in terms of design of table layout, choice of data for inclusion, ordering of data within the table, and clear and informative row/column headings and table title. Many of the visual design elements are common to those discussed for figures: keep tables free of clutter, and define abbreviations in the title or by using footnotes. In addition, don't box tables, and use horizontal lines as separators and space to separate columns.

A review of tables in published articles shows a number of common weaknesses which reduce the power of tables to contribute to the communication of the story:

- weak descriptive titles are used when a story-telling title would be appropriate (many of the points discussed in Chapter 10 on article titles apply to titles for tables as well);
- inclusion of unnecessary or redundant data (e.g. data that are not referred to in the text and do not contribute to the story, or columns of a known constant);
- inclusion of non-significant or over-precise numbers (which lead to a false sense of accuracy or clutter, respectively);
- omission of data necessary for the reader to make important calculations from experimental data (omitted from either the tables or text);
- table not arranged to highlight the most significant results;
- data not sorted to show important relationships between elements.

Tables 5.2 and 5.3 show data from a study using different methods of analyzing potassium (K) concentration in soils with different mineralogy. Table 5.3 has been modified to increase the story value of the data presented.

Table 5.2 Soil test K and mineralogy of soils (SD = Standard Deviation).

| Soil | Clay (g kg^{-1}) | Silt (g kg^{-1}) | mg K kg^{-1} soil | | |
			WS	CaCl$_2$	NaTPB
1	380	200	10	41	480
2	535	265	31	162	1208
3	410	230	15	57	583
4	434	205	19	70	652
5	485	235	27	100	932
6	610	282	50	290	1730
7	360	190	6	34	360
8	440	235	20	87	723
Mean	456.8	230.3	22.3	105.1	833.5
SD ($\pm$)	83.4	31.9	13.9	84.9	448.9

Table 5.3 Soil texture correlates with K concentration determined using three extraction methods: WS = Water Soluble, CaCl$_2$ = Calcium Chloride, NaTPB = Sodium Tetraphenyl Boron (SD = Standard Deviation).

| Soil | Clay (g kg^{-1}) | Silt (g kg^{-1}) | mg K kg^{-1} soil | | |
			WS	CaCl$_2$	NaTPB
7	360	190	6	34	360
1	380	200	10	41	480
3	410	230	15	57	583
4	434	205	19	70	652
8	440	235	20	87	723
5	485	235	27	100	932
2	535	265	31	162	1208
6	610	282	50	290	1730
Mean	457	230	22	105	834
SD ($\pm$)	83	32	14	85	449

Improvements in Table 5.3 in comparison with Table 5.2 are described below.

- The title has changed from a description to a story-telling statement of what the data show.
- Sampled soils are sorted to better highlight the gradient of soil clay content in the different soils (in Table 5.2 the soils were presented in the order in which they were collected). The soil samples could be renamed to present them in the new order.
- Mean and standard deviation values have been rounded back (which prevents presentation of false accuracy and reduces clutter).
- A small break between the individual data points and the mean values improves the visual appreciation of the gradient of soil texture and K concentrations.

> ### Task 5.2 Evaluating table design
>
> Examine the tables in your SA or another article from a journal in your field.
>
> - Are all data necessary, and are they sorted to make the main results most prominent?
> - Is the title descriptive or story-telling? Could a story-telling title be written for the table?
> - Are all numbers calculated to the correct number of significant figures and rounded to show appropriate precision?
> - Does the table have any of the weaknesses described above and how do these detract from the telling of the story?

5.4 Figure legends and table titles

Figure legends and table titles should explain what the data being presented are and highlight the key points of the part of the results story presented there. The key points of the story presented should stand alone; i.e. the reader should not need to read the rest of the text to understand them. Tables and figures which effectively and clearly communicate a part of the story make the work of reviewers easy and improve the readability of articles for all users.

Figure legends have a general form with five parts. These parts usually occur in sequence, but explanation of symbols and notation (Part 5, see below) may be interspersed in the other parts.

1 A title which summarizes what the figure is about.
2 Details of results or models shown in the figure or supplementary to the figure.
3 Additional explanation of the components of the figure, methods used, or essential details of the figure's contribution to the results story.
4 Description of the units or statistical notation included.
5 Explanation of any other symbols or notation used.

Table titles can also include all of these elements but tend to have only brief Parts 2 and 3 and not to have a Part 5.

> ### Task 5.3 Identifying parts of figure legends
>
> Read the figure legends from the Results sections of Britton-Simmons and Abbott (2008) and Kaiser et al. (2003) below and identify the parts of the figure legend described in Section 5.4.
>
> > Number of *Sargassum muticum* (a) recruits and (b) adults in field experiment plots (900 cm²). Propagule pressure is grams of reproductive tissue suspended over experimental plots at beginning of experiment. The average mass of an adult *S. muticum* (174 g) is indicated by an arrow. Data are means ± 1 SE ($n = 3$). (from Britton-Simmons & Abbott 2008, Figure 1)
>
> *(Continued)*

Task 5.3 (*Continued*)

Uptake of Fe(II) by GmDmt1 in yeast.

(a) Influx of $^{55}Fe^{2+}$ into yeast cells transformed with GmDmt1;1, *fet3fet4*cells were transformed with GmDmt1;1-pFL61 or pFL61 and then incubated with 1 μM $^{55}FeCl_3$(pH 5.5) for 5- and 10-min periods. Data presented are means ± SE of ^{55}Fe uptake between 5 and 10 min from three separate experiments (each performed in triplicate).

(b) Concentration dependence of ^{55}Fe influx into *fet3fet4*cells transformed with GmDmt1;1-pFL61 or pFL61. Data presented are means ± SE of ^{55}Fe uptake to over 5 min (n = 3). The curve was obtained by direct fit to the Michaelis-Menten equation. Estimated K_M and V_{MAX} for GmDmt1;1 were 6.4 ± 1.1 μM Fe(III) and 0.72 ± 0.08 nM Fe(III) $min^{-1}mg^{-1}$ protein, respectively.

(c) Effect of other divalent cations on uptake of $^{55}Fe^{2+}$ into *fet3fet4*cells transformed with pFL61-GmDMT1;1. Data presented are means ± SE of ^{55}Fe (10 μM) uptake over 10 min in the presence and absence of 100 μM unlabelled Fe^{2+}, Cu^{2+}, Zn^{2+} and Mn^{2+}. (from Kaiser et al. 2003, Figure 5)

Check your answers in the Answer pages.

CHAPTER 6

Writing about results

In writing sentences about their results, effective authors highlight the main points only. Published advice from editors and researchers indicates that it is important that authors *do not* repeat in words all the results from the tables or figures. This advice often suggests that authors should only write sentences about the most important findings, especially the ones that will form part of the focus of the Discussion section.

Results are sometimes presented separately from the Discussion and sometimes combined in a single Results and discussion section. Check in the Instructions to Contributors for the journal you are targeting to see which format they prefer, or examine a selection of articles if the Instructions to Contributors are not sufficiently explicit.

If the separate style is used, it is generally important to confine any comments in the Results section to saying what the numbers show, without comparing them with other research, or suggesting explanations. However, authors sometimes include comparisons with previous work in the Results section where the point being made relates to a component of the results that will not be discussed in detail in the Discussion. For an example, see the first PEA, Kaiser et al. (2003), p. 126, column 2, line 7 and following.

In general, keeping Results and Discussion sections separate is more common.

6.1 Functions of results sentences

The text of a Results section typically

- highlights the important findings;
- locates the figure(s) or table(s) where the results can be found; and
- comments on (but does not discuss) the results.

Elements that highlight and locate are sometimes combined in the same sentence, and sometimes appear in separate sentences.

Examples of combined highlight + location styles

Measurements of root length density (Figure 3) revealed that the majority of roots of both cultivars were found in the upper substrate layers.

The response of lucerne root growth to manganese rate and depth treatments was similar to that of shoots (Figure 2).

Writing Scientific Research Articles: Strategy and Steps, 1st edition. By M. Cargill and P. O'Connor. Published 2009 by Blackwell Publishing, ISBN 978-1-4051-8619-3 (pb) and 978-1-4051-9335-1 (hb)

Figure 17 shows the average number of visits per bird.

Note the different verb tenses used in the two styles.

Task 6.1 Separate location sentences in Results sections

First skim (read quickly) the Results section of your selected PEA. Count how many instances of separate location sentences you find. Why do you think the authors chose to write their Results section as they did? Check your answers in the Answer pages.

Now do the same exercise for your SA. Discuss your findings with a colleague, if appropriate.

6.2 Verb tense in Results sections

Task 6.2 Verb usage in Results sections

1 Read the extract from a Results section below and identify which **verb tenses/verb forms** are represented by the underlined words in each sentence (present, past, or modal verb). Can you think of a reason for the use of different tenses in different sentences? (N.B. The past participles used as adjectives in the passage have not been underlined, only the finite verbs.)

Antibodies <u>were raised</u> in rabbits against the N-terminal 73 amino acids of GmDmt1;1 (Figure 1c). This antiserum <u>was used</u> in Western blot analysis of 4-week-old total soluble nodule proteins, nodule microsomes, PBS proteins and PBM, isolated from purified symbiosomes. The anti GmDMT1 antiserum <u>identified</u> a 67-kDa protein on the PBM-enriched nodule protein fraction (Figure 3a), but <u>did not cross-react</u> with soluble nodule proteins, PBS proteins or nodule microsomes (Figure 3a). Replicate Western blots incubated with pre-immune serum (Figure 3b) <u>did not cross-react</u> with the soybean nodule tissue examined. The protein identified on the PBM-enriched protein fraction <u>is</u> approximately 10 kDa larger than that predicted by the amino acid sequence of GmDmt1. The increase in size <u>may be related</u> to extensive post-translational modification (e.g. glycosylation) of GmDmt1, as it <u>occurs</u> in other systems. (Kaiser et al. 2003)

2 Summarize your findings using the following sentence starters:

In Results sections, the past tense is used to talk about ...
The present tense is used in sentences that ...
Modal verbs are used to ...

Compare your answers with the points below.

Common use of tense in Results sections

- Past tense (either active or passive voice) is used when the sentence focuses on the completed study: what was done and found.

- Present tense is used:
  - to describe an "always true" situation; and
  - when the sentence focuses on the document, which will always be there. N.B. Although there are no examples of this usage in the above paragraph from Kaiser et al. (2003), here is an example from McNeill et al. (1997):

The effect of urea concentration on the fed leaf and shoot growth in subterranean clover is summarised in Table 1.

- Modal verbs (e.g. *may* and *could*) may be used in comments, especially in *that* clauses. (See Chapter 9 for more details about modal verb use in research writing.)

Task 6.3 Analysing your SA Results section for verb usage

Choose one subsection of the Results section in your SA. Answer the following questions and discuss your findings with a colleague.

- For each verb in the subsection, why do you think the author(s) chose to use the tense they did?
- Do the authors use tenses in the ways discussed in the section above? If not, what reasons can you suggest?

If you find many instances where the tense usage differs from the guidelines given above, we suggest that you look at two or three other papers from your field and check the tense usage in their Results section as well. If you discover patterns that differ from our guidelines, congratulations! Make a note of your findings to guide your own future use.

Hint: Example papers from your own discipline provide the most accurate guidelines for you.

It is probably not possible to write a book that presents accurately the writing conventions of every different subfield of science. Rather than aiming to provide all the answers, we have set out to give you tools and questions to use in analysing example articles from your own research area. We want you always to check what we suggest against these examples and in this way to refine the guidelines we give, so they are as accurate as possible for the articles you need to write, in order to submit to journals relevant to your field. We believe this comparison process is a valuable component of the descriptive and discovery-based method for learning about research article writing that we present in this book.

Task 6.4 Drafting your own Results section

Begin to draft a Results section for your own paper (OA), writing about the tables or figures you have worked on previously.

CHAPTER 7

The Methods section

7.1 Purpose of the Methods section

Traditionally, students are taught that the Methods section provides the information needed for another competent scientist to repeat the work. In your experience of reading papers, is this what you find? Many participants in workshops we have conducted report that they have had problems in replicating what authors have done in their published studies even after reading the Methods section thoroughly.

Another way to think about the goal of the Methods section is that it establishes credibility for the results and should therefore provide enough information about how the work was done for readers to evaluate the results; i.e. to decide for themselves whether the results actually mean what the author claims they mean. Referees are likely to look in this section for evidence to answer the question: Do the methods and the treatment of results conform to acceptable scientific standards?

A short note on the naming of this section of a research article is in order here. As you have seen from your analysis of the PEAs in Chapter 2, practice varies. Alternatives include Methods, Materials and methods, and Experimental procedures. For the sake of simplicity, we use the term Methods throughout this chapter.

It is generally accepted that methods that have been published previously can be cited and need not be described in detail, unless changes have been made to the published procedures. However, if the previous publication is not readily available to your international audience (e.g. the original journal is written in a language other than English), it is recommended that you give the details in your paper, as well as the citation to the original source. Include the language of its publication in brackets in the reference list, if appropriate. Any novel method should be described in full.

7.2 Organizing Methods sections

If a goal of the Methods section is to help readers evaluate the findings presented in the Results section, then the author needs to make it clear how the two sections relate to each other, and the Methods usually comes *before* the Results. Two strategies can help with showing the connections.

Writing Scientific Research Articles: Strategy and Steps, 1st edition. By M. Cargill and P. O'Connor. Published 2009 by Blackwell Publishing, ISBN 978-1-4051-8619-3 (pb) and 978-1-4051-9335-1 (hb)

- Strategy 1 Use identical or similar subheadings in the Methods and the Results sections.
- Strategy 2 Use introductory phrases or sentences in the Methods that relate to the aims, e.g.

> To generate an antibody to GmDmt1;1, a 236-bp DNA fragment coding for 70 N-terminal amino acids was amplified using the PCR,...

An additional strategy to clarify the logic of the Methods section is to use the first sentence of a new paragraph to introduce what you will be talking about and relate it to what has gone before. In the example below, *disturbance treatment* refers to a concept that has been mentioned previously, and the sentence introduces the reader effectively to the content of the paragraph to follow (Britton-Simmons & Abbott 2008, p. 137, paragraph 2):

> The disturbance treatment had two levels: control and disturbed. Control plots were ...

Task 7.1 Materials and methods organization

Look at the Methods section of your selected PEA and answer the questions.

1 What subheadings are used in the section?
2 How do the subheadings relate to

   i the end of the Introduction?
   ii the subheadings in the Results section?

3 Is the section easy for you to follow? Why? Or why not?

Compare your answers with our suggestions in the Answer pages.
   Now, repeat the task for your SA, and discuss your findings with a colleague or teacher if appropriate.

Task 7.2 Planning your Methods section

For your OA, which elements do you plan to include in the Methods section, and in what order?

7.3 Use of passive and active verbs

Researchers commonly write about materials and methods in the *passive voice*: that is, using passive voice verbs. These verb forms emphasize the action, and remove emphasis from the doer of the action, but they often use more words than the corresponding active voice verbs. Many books written to advise researchers about improving their writing recommend that authors avoid the passive, and use active verbs as much as possible, because this makes the writing more direct and less wordy. We agree that the passive is often over-used in science writing in general.

However, we suggest that the choice is not always a simple one, especially in Methods sections, and in this section we will do the following things:

- refresh your memory on the difference between active and passive verb forms;
- consider reasons why an author may wish to choose a passive verb; and
- present some guidelines for avoiding common problems with passive verb use.

Active and passive verb forms

When we use an active verb, the grammatical subject of the verb (the answer to who or what in front of the verb) actually does the action indicated by the verb. For example:

subject	+	active verb	+	object
The dog		**bit**		**the man.**

With a passive verb, the grammatical subject does not do the action of the verb (the biting, in this case). For example:

subject	+	passive verb	+	agent
The man		**was bitten**		**by the dog.**

The agent is often omitted in passive sentences, which is why this form is popular when the action is more important than the actor, as in many experimental procedures.

Figure 7.1 summarizes the difference between the two sentence constructions.

If authors of research articles are comfortable with using active voice sentences with "we" as the subject, as in the example in Figure 7.1, then it is relatively easy to avoid the passive voice, even in Methods sections. However, many authors are not comfortable with this usage, or do not like the repetitive sound of many "we" sentences together, and many passive verbs can still be found in science writing.

Formation of passive voice verbs requires an auxiliary – i.e. a part of the verb *to be* (*was* is used in the example above) – plus the past participle of a verb (*bitten* in the example above). Remember, only a transitive verb, a verb that has an object (indicated in dictionaries as vt.), can have a passive form.

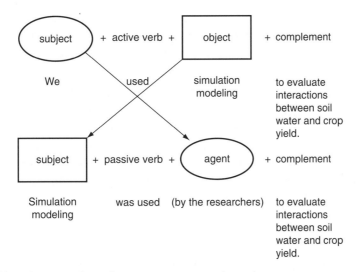

Fig. 7.1 Changing an active voice sentence to a passive voice sentence.

Task 7.3 Active/passive sentences

Find one passive sentence from the Methods section of your selected PEA, and rewrite it in the active voice. Then find a sentence in the active voice that uses a transitive verb, and rewrite it in the passive voice. We provide some sample answers from each article in the Answer pages.

Factors influencing the choice of an active or passive verb

First, does the reader need to know who or what carried out the action? If this information is unimportant, you may choose to use a passive verb. Consider the following example.

> The researchers collected data from all sites weekly.

It is not important *who* collected the data, so the sentence may be better in the passive:

> Data* were collected weekly from all sites.

Second, does it sound repetitive (or immodest) to use a personal pronoun subject? For example:

> We calculated least significant differences (l.s.d.) to compare means.

This may sound more appropriate in the passive:

> Least significant differences (l.s.d.) were calculated to compare means.

Note the following points in relation to active/passive choice.

- The need to avoid repetition can explain the almost complete absence of active voice sentences in the Experimental procedures section of the PEA by Kaiser et al. (2003) (Chapter 18): in the active, the subject of nearly every sentence would be "we".
- If you are working in a discipline where single-authored papers are common, you will need to check in a range of example papers whether it is appropriate to use "I"; in our experience this usage is quite rare in science writing, especially in Methods sections.
- Does it help the information flow to choose either the active or passive voice?

In English sentences, effective writers generally connect their sentences to each other by putting old information, which the reader already knows something about, before new information (see section 8.8 for a fuller explanation of this linking strategy). Sometimes writers may choose a passive verb so that they can use this strategy. In the example below, the old information is in italic, and the active and passive verbs are identified.

---

*N.B. Data is a plural word of Latin origin, and it is still common for editors to require its use with plural verb forms. However, this convention is in the process of changing and you are likely to see it used both ways: the data show, and the data shows.

We used [ACTIVE] the results of these analyses to inform the construction of mechanistic candidate functions for the relationship between propagule input, space availability and recruitment. *These candidate functions* were compared [PASSIVE] using differences in the Akaike information criteria (AIC differences; Burnham and Anderson 2002). We then used model averaging [ACTIVE] .... (Britton-Simmons & Abbott 2008, p. 137)

Common problems with writing passive sentences

There is one common problem with writing passive sentences that makes them unwieldy and difficult for your reader to follow. In order to make your writing easier to understand, take particular care *not* to write sentences with very long subjects and a short passive verb right at the end. For example:

   ✗   Wheat and barley, collected from the Virginia field site, as well as sorghum and millet, collected at Loxton, were used.

Instead, try to get both the subject and the verb within the first nine words of the sentence, and make sure any list of items is at the end of the sentence, as in the following example.

   ✓   Four cereals were used: wheat and barley, collected from the Virginia field site; and sorghum and millet, collected at Loxton.

N.B. This improved example demonstrates a very effective sentence structure for writing lists in English. A short introduction clause (which could be a sentence on its own) is followed by a colon (:) to introduce the list. Because the two items in the list have internal commas, the items themselves are separated with a semicolon (;). This use of punctuation makes it very clear which parts of the sentence belong together, and which are separated.

Task 7.4 Top-heavy passive sentences

1 Here is another example of a top-heavy sentence, with a very long subject followed by a short passive verb near the end. Rewrite the sentence to make it easier for a reader to understand.

Actual evapotranspiration (T) for each crop, defined as the amount of precipitation for the period between sowing and harvesting the particular crop plus or minus the change in soil water storage in the 2m soil profile, was computed by the soil water balance equation (Xin, 1986; Zhu and Niu, 1987).
From Li et al. (2000).

Check your answer in the Answer pages.

2 Select one subsection of the Methods in your SA and check whether the authors have avoided this problem. Can you find any sentences that are difficult to follow? How could you improve them? Discuss your findings with a colleague.

Table 7.1 Abbreviating passive sentences to avoid excessive repetition.

Original sentence	Possible abbreviation
The data were collected and they were analysed using...	The data were collected and analysed using...
The data were collected and correlations were calculated...	The data were collected and correlations calculated...
The data which were collected were analysed using...	The data collected were analysed using...

Abbreviating passive sentences to avoid sounding repetitive

You may find it useful to abbreviate passive sentences, as shown in Table 7.1.

Task 7.5 Revising your own Methods section

Use what you have learned to improve your draft of the Methods section of your own paper (OA).

CHAPTER 8

The Introduction

As your primary reading audience of editor and referees will probably start reading at the Introduction, an effective Introduction is particularly important. Referees are likely to look here for evidence to answer the following questions.

1 Is the contribution new?
2 Is the contribution significant?
3 Is it suitable for publication in the journal?

8.1 Five stages to a compelling Introduction

Applied linguistics researchers have identified five main stages that commonly appear in research article Introductions (Figure 8.1). These stages have been identified through analyzing many published articles, and interesting variations have been found across different subdisciplines of science. However, for our purposes in this book, the five broad stages give us a useful framework that is flexible enough to be applicable in most contexts. But please remember that they do not represent a recipe to be followed unreflectively; rather, they provide a pattern for you to test on papers in your own field, and to refine into a useful tool for your own use.

These stages do not always occur strictly in the order given in Figure 8.1, and some may be repeated within a given Introduction. For example Stage 2/Stage 3 sequences often recur when an author wants to justify specific aspects or components of a study. To help you see what we mean by these stages, we first ask you to read the article introduction presented in Table 8.1 and consider our identification of the stages and their locations.

Task 8.1 Introduction stages

Read the introduction of your selected PEA, decide if all stages are present, and mark where each one begins and ends. (Remember that it is possible that stages may be repeated or come in a different order to that suggested in Figure 8.1.)
Compare your findings with our suggestions in the Answer pages.
Now, do the same for your own SA. Discuss your findings with a colleague or teacher if appropriate.

Writing Scientific Research Articles: Strategy and Steps, 1st edition. By M. Cargill and P. O'Connor. Published 2009 by Blackwell Publishing, ISBN 978-1-4051-8619-3 (pb) and 978-1-4051-9335-1 (hb)

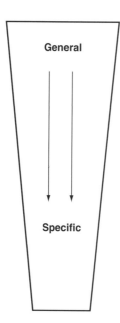

1. Statements about the field of research to provide the reader with a setting or context for the problem to be investigated and to claim its centrality or importance.

2. More specific statements about the aspects of the problem already studied by other researchers, laying a foundation of information already known.

3. Statements that indicate the need for more investigation, creating a gap or research niche for the present study to fill.

4. Statements giving the purpose/ objectives of the writer's study or outlining its main activity or findings.

5. Optional statement(s) that give a positive value or justification for carrying out the study.

Fig. 8.1 Five stages of an Introduction to a science research article (after Weissberg & Buker 1990).

Table 8.1 Identification of stages in the Introduction to "Use of *in situ* ^{15}N-labelling to estimate the total below-ground nitrogen of pasture legumes in intact soil-plant systems" (McNeill et al. 1997).

Extract	Stage
Current estimates of the below-ground production of N by pasture legumes are scarce and rely mainly on data from harvested macro-roots (Burton 1976; Reeves 1984) with little account taken of fine root material or soluble root N leached by root washing. Sampling to obtain the entire root biomass is extremely difficult (Sauerbeck and Johnen 1977) since many roots, particularly those of pasture species (Ellis and Barnes 1973), are fragile and too fine to be recovered by wet sieving. Furthermore, the interface between the root and the soil is not easy to determine and legume derived N will exist not only as live intact root but in a variety of other forms, often termed rhizodeposits (Whipps 1990). An approach is accordingly required which enables *in situ* labelling of N in the legume root system under undisturbed conditions coupled with subsequent recovery and measurement of that legume N in all of the inter-related below-ground fractions.	Stage 1 Stage 3 in "scarce" and "little account" Stage 1 Stage 3 (broad gap)
Sophisticated techniques exist to label roots with ^{15}N via exposure of shoots to an atmosphere containing labelled NH$_3$ (Porter *et al.* 1972; Janzen and Bruinsma 1989) but such techniques would not be suitable for labelling a pasture legume within a mixed sward. Labelled N$_2$ atmospheres (Warembourg *et al.* 1982; McNeill *et al.* 1994) have been used to label specifically the legume component of a mixed sward via N$_2$ fixation in nodules. However, these techniques require complex and expensive enclosure equipment, which limits replication and cannot be easily applied to field situations; furthermore, non-symbiotic N$_2$ fixation of label may occur in some soils and complicate the interpretation of fate of below-ground legume N.	Stage 2 Stage 3 Stage 2 Stage 3

(Continued)

Table 8.1 (*Continued*)

Extract	Stage
The split-root technique has also been used to introduce ^{15}N directly into plants by exposing one isolated portion of the root system to ^{15}N either in solution or soil (Sawatsky and Soper 1991; Jensen 1996), but this necessitates some degree of disturbance of the natural system. Foliar feeding does not disturb the system and has the additional advantage that shoots tolerate higher concentrations of N than roots (Wittwer *et al.* 1963). Spray application of ^{15}N-labelled urea has been successfully used to label legumes *in situ* under field conditions (Zebarth *et al.* 1991) but runoff of ^{15}N-labelled solutions from foliage to the soil will complicate interpretation of root-soil dynamics. Russell and Fillery (1996), using a stem-feeding technique, have shown that *in situ* ^{15}N-labelling of lupin plants growing in soil cores enabled total below-ground N to be estimated under relatively undisturbed conditions, but they indicated that the technique was not adaptable to all plants, particularly pasture species. Feeding of individual leaves with a solution containing ^{15}N is a technique that has been widely used for physiological studies in wheat (Palta *et al.* 1991) and legumes (Oghoghorie and Pate 1972; Pate 1973). The potential of the technique for investigating soil-plant N dynamics was noted as long as 10 years ago by Ledgard *et al.* (1985) following the use of ^{15}N leaf-feeding in a study of N transfer from legume to associated grass. The experiments reported here were designed (*i*) to assess the use of a simple ^{15}N leaf-feeding technique specifically to label *in situ* the roots of subterranean clover and serradella growing in soil, and (*ii*) to obtain quantitative estimates of total below-ground N accretion by these pasture legumes.	Stage 2 Stage 3 Stage 2 Stage 3 Stage 2 Stage 3 Stage 2 (Stage 3 implicit in "potential") Stage 4 (aims of the present study)

8.2 Stage 1: Locating your project within an existing field of scientific research

Constructing the right setting for your paper

In Stage 1, authors mostly begin with broad statements that would generally be accepted as fact by the members of their reading audience. The present tense is often used for this kind of statement because one function of the present tense in English is expressing information perceived as always true. Sentences written in the present perfect tense are also common in Stage 1, expressing what has been found over an extended period in the past and up to the present. These statements may or may not include references, depending on the field and the topic of the paper.

Task 8.2 Introduction Stage 1 analysis

1 Check the first paragraphs of the Introductions of the two PEAs and complete Table 8.2. Then check your answers with our suggestions in the Answer pages.

(Continued)

Task 8.2 (*Continued*)

2 Now repeat the exercise for your SA, compare your findings with those for the PEAs, and discuss any differences with a colleague or teacher, if appropriate.

Table 8.2 Task 8.2: Introduction Stage 1 analysis.

Question	Kaiser et al. (2003)	Britton-Simmons and Abbott (2008)
Are some sentences written in the present tense? How many?		
Are some sentences written in the present perfect tense? How many?		
Which tense is used more? Why do you think this is the case?		
How many sentences contain references?		
What kinds of sentences do not have references?		

Authors then seek to move their readers smoothly from these broad, general statements towards one sub-area of the field, and then to the authors' own particular topic. One way to think about this is to begin in a selected country and imagine you are moving from that country (the broad area where the Introduction begins) and zooming in on a province in that country, and finally focusing on a particular city, which represents the topic area of research to be presented in the paper.

Task 8.3 Country to city in Stage 1

1 Look at the Introduction of your selected PEA. What is the country? The province? The city?

Check your answers against our suggestions in the Answer pages.

2 Now do the same task for the Introduction to the SA you are analyzing. Country? Province? City?

3 Now try to suggest these three features for your OA. Remember, your "city" is not your purpose for conducting the study, but rather the specific topic area for your paper. Country? Province? City?

Writers move their readers through these steps by linking their sentences through the positioning of *old* and *new information*. Old information is any information that the reader already knows; it is placed towards the beginning of sentences. New information comes towards the end of sentences. (This convention is very important for improving flow in all forms of technical writing.) See Task 8.4.

8.3 Using references in Stages 2 and 3

In Stages 2 and 3 of an Introduction (see Figure 8.1) authors use selected literature from their field to justify their study and construct a gap or niche for

> ## Task 8.4 Identifying old or given information
>
> Look at the extract from the Introduction in Kaiser et al. (2003) (see Chapter 18) below and underline the words that represent or refer to *old information* (information the reader already knows about, also called *given information*).
>
> Legumes form symbiotic associations with N_2-fixing soil-borne bacteria of the *Rhizobium* family. The symbiosis begins when compatible bacteria invade legume root hairs, signalling the division of inner cortical root cells and the formation of a nodule. Invading bacteria migrate to the developing nodule by way of an 'infection thread', comprised of an invaginated cell wall. In the inner cortex, bacteria are released into the cell cytosol, enveloped in a modified plasma membrane (the peribacteroid membrane (PBM)), to form an organelle-like structure called the symbiosome, which consists of bacteroid, PBM and the intervening peribacteroid space (PBS; Whitehead and Day, 1997). The bacteria, subsequently, differentiate into the N_2-fixing bacteroid form. The symbiosis allows the access of legumes to atmospheric N_2, which is reduced to NH_4^+ by the bacteroid enzyme nitrogenase. In exchange for reduced N, the plant provides carbon to the nodules to support bacterial respiration, a low-oxygen environment in the nodule suitable for bacteroid nitrogenase activity, and all the essential nutritional elements necessary for bacteroid activity. Consequently, nutrient transport across the PBM is an important control mechanism in the promotion and regulation of the symbiosis.
>
> Check your answers in the Answer pages.

their own work. They write sentences supported by *references* to the literature they have selected. In this context, the term literature refers to all the published research articles, review articles, and books in a given field. The term also includes information published on websites that have been peer-reviewed or belong to organizations with appropriate scientific reputations.

Referencing: how to do it and why you need to

References to other published studies, also known as citations or in-text citations, can be used in all stages of the Introduction, as you have seen in the samples we have looked at. They appear in the text either as a surname and year in brackets, e.g. (McNeill 2000), or as a number, e.g. [7]. The details of the presentation depend on the style stipulated by the journal. Check the Instructions to Contributors of your target journal for the necessary information on referencing style. These references *refer* to the list of references at the end of the paper, where the full publication details are written.

Citations are particularly vital in showing that you know clearly the work that has been conducted by others in your city area (see Task 8.3 above), and therefore what has not been done and needs to be done: the gap that your study will fill. This function is carried out in Stages 2 and 3. What you are required to do here is, in effect, to construct an argument which justifies your own study and shows why and how it is important.

Using citation to develop your own argument

Below are examples of parts of paragraphs using three different citation methods (the references cited have been invented for demonstration purposes only). These methods can be called information prominent, where the focus of the sentence is only on the information being presented; author prominent, where the name of the author of the information is given prominence in the sentence; and weak author prominent, where the ideas of author(s) are given prominence, but author names do not appear in the main part of the sentence. Observe how the different methods contribute to the way in which the writer's argument is developed. (N.B. For this section, the term *author* is used for the author of a published paper that is being cited; the term *writer* is used to refer to the person writing the text that cites the author's work.)

Information prominent citation

Shrinking markets are also evident in other areas.* The wool industry is experiencing difficulties related to falling demand worldwide since the development of high-quality synthetic fibres (Smith 2000).

This is the default style in many areas of science and is the only style used in the Introductions of the two PEAs. However, there are two other options that should also be part of a writer's repertoire, for use when appropriate.

Author prominent citation style 1

Shrinking markets are also evident in other areas. As Smith (2000) pointed out, the wool industry is experiencing difficulties related to falling demand worldwide since the development of high-quality synthetic fibres.

This style gives more option to show the writer's view of the cited fact. In this case, it shows that the writer (you!) agrees with Smith.

or ### Author prominent citation style 2

Shrinking markets are also evident in other areas. Smith (2000) argued that the wool industry was experiencing difficulties related to falling demand worldwide since the development of high-quality synthetic fibres. However, Jones et al. (2004) found that industry difficulties were more related to quality of supply than to demand issues. It is clear that considerable disagreement exists about the underlying sources of these problems.

This style also allows the use of verbs such as argued, which give the reader advance notice that a however or some other contrast may be coming, and indicate that what is being cited is not necessarily accepted as correct by you, the writer. However, there is a danger attached to the author prominent style. If it is over-used, it can make the text sound like a list, rather than a logically constructed argument. We recommend that you use this style sparingly, perhaps when you are approaching the specifics of the gap your study will address. It is also useful to pay close attention to the papers you read in your own field, to check how often, if at all, this style appears.

or ### Weak author prominent citation

Several authors have reported that the wool industry is experiencing difficulties related to falling demand since the development of high-quality synthetic fibres (Smith 2000, Wilson 2003, Nguyen 2005). For example, Smith (2000) highlighted ...

*This first sentence is a "topic sentence" for the paragraph: its function here is to form a link to the previous paragraph (which discussed shrinking markets), and to alert the reader to the topic of the current paragraph. Topic sentences are an effective way of creating logical flow in science writing.

Task 8.5 Citation styles in an authentic example

Read the Introduction extract presented in Table 8.3 and observe how the different citation styles are used.

Table 8.3 Use of different citation styles in a segment of the Introduction from McNeill et al. (1997).

Introduction text	Citation style
Foliar feeding does not disturb the system and has the additional advantage that shoots tolerate higher concentrations of N than roots (Wittwer et al. 1963).	Information prominent
Spray application of ^{15}N-labelled urea has been successfully used to label legumes *in situ* under field conditions (Zebarth et al. 1991) but runoff of ^{15}N-labelled solutions from foliage to the soil will complicate interpretation of root-soil dynamics.	Information prominent / Writer's evaluation statement
Russell and Fillery (1996), using a stem-feeding technique, have shown that *in situ* ^{15}N-labelling of lupin plants growing in soil cores enabled total below-ground N to be estimated under relatively undisturbed conditions, but they indicated that the technique was not adaptable to all plants, particularly pasture species.	Author prominent
Feeding of individual leaves with a solution containing ^{15}N is a technique that has been widely used for physiological studies in wheat (Palta et al. 1991) and legumes (Oghoghorie and Pate 1972; Pate 1973).	Information prominent
The potential of the technique for investigating soil-plant N dynamics was noted as long as 10 years ago by Ledgard et al. (1985) following the use of ^{15}N leaf-feeding in a study of N transfer from legume to associated grass.	Author prominent, but using the passive voice so that the link (technique) can come first in the sentence as old information.

This method has a general reference to authors in the subject and then more than one reference in the brackets. It is followed here by an author prominent citation. This style can be useful as a topic sentence when beginning a new subtopic or line of argument. Note that this style requires the use of the present perfect tense (*have reported*).

Writers choose their citation method to fit with the way their paragraph is advancing their argument.

Citing when you cannot obtain the original reference

Editors usually require that writers cite only those papers that they have actually read. However, if you cannot obtain the original article and are therefore obliged to rely on another author's interpretation of a fact or finding you want to cite, you may use the following form of secondary citation in-text.

[The finding or fact you want to cite] (Smith 1962, cited in Jones 2002).

In such cases, only Jones (2002) appears in the reference list.

Another important reason to pay careful attention to referencing is to avoid plagiarizing other people's work unintentionally. Plagiarism is using data, ideas, or words that originated in work by another person without appropriately acknowledging their source. It is generally regarded as a form of cheating in academic and publishing contexts, and papers will be rejected if plagiarism is detected. Incomplete citation also prevents your gaining credit for knowing the work of other researchers in the field. Effective and inclusive citation helps you present yourself as a knowledgeable member of the research community, which can be important in terms of the impression you make on referees evaluating your manuscripts. It also allows others to benefit from the sources of information you have used.

Avoiding plagiarism requires writers to do two things: to be aware of the kinds of situations where inadvertent plagiarism is likely to occur; and to develop effective note-taking practices to ensure they remain aware of the status of their notes as they convert them into sentences in a paper for submission.

Task 8.6 Identifying plagiarism

Below are two versions of the same information, adapted from the Introduction in McNeill et al. (1997). In version 2, identify where the writer has plagiarized by writing in his or her own voice ideas that originated in another document (as demonstrated in version 1).

Version 1 Russell and Fillery (1996), using a stem-feeding technique, have shown that *in situ* ^{15}N-labelling of lupin plants growing in soil cores enabled total below-ground N to be estimated under relatively undisturbed conditions, but they indicated that the technique was not adaptable to all plants, particularly pasture species.

Version 2 Russell and Fillery (1996), using a stem-feeding technique, have shown that *in situ* ^{15}N-labelling of lupin plants growing in soil cores enabled total below-ground N to be estimated under relatively undisturbed conditions. However, this technique is not adaptable to all plants, particularly pasture species.

Check your answers in the Answer pages.

The important thing to watch for is that it is clear to your reader whether the idea or fact you are using in each and every sentence is your own, or has come from the work of another person. If it comes from someone else's work, cite them! It is possible that the person whose idea it originally was will be a referee of your paper, and they will be sure to notice the problem. In any case, the referees will know the literature well, so it is very important to be accurate in your citation practices.

Remember also that direct quotations using quotation marks or inverted commas ("...") are extremely rare in science writing. This means that authors need to paraphrase sentences that appear in the work of other authors, rather than copying them verbatim. However, remember also that you can expand your repertoire of sentence structures by removing the content (most often the noun phrases, indicated by NP in the example below) from sentences that appeal to you and re-using the shell (or sentence template) for your own content. For example, from the sentence in Task 8.6, version 1, you could reuse this shell:

> [Authors], using [NP1], have shown that [NP2] enabled [NP3] to be estimated under [adjective] conditions, but they indicated that the technique was not adaptable to all [NP4], particularly [NP5].

See Chapter 17 for more details of this approach.

8.5 Indicating the gap or research niche

This is Stage 3 of an Introduction (see Figure 8.1), and it can be written in a multitude of ways. As discussed previously, authors often present a broad gap early in the Introduction, and a more specific one close to the end. Examples include the following, taken from Britton-Simmons and Abbott (2008) (see Chapter 19):

> However, understanding how these processes interact to regulate invasions remains a major challenge in ecology.

> Despite its acknowledged importance, propagule pressure has rarely been manipulated experimentally and the interaction of propagule pressure with other processes that regulate invasion success is not well understood.

> It is presently unclear how different disturbance agents influence long-term patterns of invasion.

It is common to find so-called signal words that indicate that a Stage 3 statement is being made. In the examples above such signal words include *however*, *remains a major challenge*, *rarely*, *not well understood*, and *presently unclear*.

Task 8.7 Signal words for the research gap or niche

Reread the Introductions from McNeill et al. (1997) (see Table 8.1) and your selected PEA, and identify the signal words that indicate a gap is being described. List them and then check the list against our suggestions in the Answer pages.

Task 8.8 Drafting your own Introduction: Stage 3

Begin to draft Stage 3 for the Introduction of your own paper, if appropriate.

8.6 Stage 4: The statement of purpose or main activity

At the end of the Introduction authors set up the readers' expectations of the rest of the paper: they tell them what they can expect to learn about the research being presented. As indicated in Figure 8.1, Stage 4 of the Introduction is generally in the form of the aim or purpose of the study to be reported, or the principal activity or finding of the study. Authors have considerable flexibility in choosing how they will word their Stage 4, and it can be instructive to pay attention to how this is done in each paper that you read for your research. You may like to keep a list of possible wordings, to help when you come to the writing of your own papers.

Task 8.9 Stage 4 sentence templates

Identify the Stage 4 in the Introduction in McNeill et al. (1997), presented in Table 8.1, and in your selected PEA. We have provided a shell, or sentence template, from each one in the Answer pages.

Task 8.10 Drafting your own Introduction: Stage 4

Draft a Stage 4 for the Introduction of your own paper, if appropriate. Write it so that it runs smoothly on from your Stage 3 gap statement, to form the closing part of your Introduction. Make sure that all the keywords in your title have been used in these sentences, to meet the expectations you set up for your readers when they read the title.

8.7 Suggested process for drafting an Introduction

Here is a summary of a process for drafting an Introduction. It is useful *after* you have made the key decisions about the results you will include in the paper, and what they mean for the audience who will read the paper.

1 Begin with Stage 4. Write an aim statement, or a statement describing what the paper sets out to do. It is usually the easiest part of the Introduction to write. It will appear in the final paragraph of the Introduction, but it is useful to write it early in the drafting process.

2 Draft Stage 3 next: the gap or need for further work. As we have seen in the previous sections, there may be one or more sub-gaps at different places in your Introduction, as well as a Stage 3 statement that leads into Stage 4. Consider beginning your Stage 3 sentences with words such as *however* or *although*, and incorporating words indicating a need for more research, such as *little informa-tion, few studies, unclear,* or *needs further investigation.*

3 Then think about how to begin Stage 1, the setting. Think about your intended audience and their interests and background knowledge, and the ideas you have highlighted in your title. Try to begin with words and concepts that will immediately grab the attention of your intended readers.

4 Next arrange the information you have collected from the literature into Stage 2. This is a very important part and you will probably need quite a bit of time to write it. You may need to do some more searching of the literature, to make sure you have done the best possible job of finding the relevant work in the area and the most recent studies.

5 Combine the stages into a coherent Introduction. You may need to add additional sentences providing background, and/or to rearrange sentences or sections to get the best possible logical development. Section 8.8 focuses on strategies for revising your Introduction to enhance the logical flow of the writing, once you are happy with the content you have included.

8.8 Editing for logical flow

In English writing, the responsibility rests with the writer to ensure that the reader recognizes the logical flow of the argument being presented. This is not the case in all languages! However, even for writers with English as a first language, the strategies for achieving this goal in their writing are often not obvious. We suggest some important strategies in the following sections. We have mentioned several of these previously in the book, but this section brings them together into a coherent set and provides you with some practice in improving poor examples.

Strategy 1: Always introduce ideas

Use informative titles, subheadings and introduction sections to set up expectations in your readers.

> A key to effective scientific and technical communication in English is to set up expectations in your reader's mind, and then meet these expectations as soon as possible.

Make the wording of your subheadings, if your target journal uses them, a part of the process of telling your reader what to expect next, in much the same way that the paper's title alerts them to the main message of the paper as a whole. In paragraphs, use the first sentence as a topic sentence to orient your readers to the main point or purpose of the paragraph. Topic sentences can also be used to link the upcoming paragraph to the one that precedes it; see Task 18.11.

Strategy 2: Move from general information to more specific information

Readers of English text expect that they will read general information about any topic or point first, before encountering details, examples, or other more specific information.

Consider the following sample paragraph and decide whether it meets the requirement to move from the general to the particular. Alternatively, is there a sentence that seems to be too general late in the paragraph? (Sentences are numbered to make it easier to refer to them later.)

> [1]Pleuropneumonia (APP) can present as a dramatic clinical disease or as a chronic, production limiting disease in pig herds. [2]A sudden increase in the number of sick and coughing pigs and a sharp rise in mortalities among grower/finisher pigs may herald an outbreak of APP in a herd. [3]On the other hand, signs may be limited to a drop in growth rate and an increase in grade two pleurisy lesions in slaughter pigs. [4]The disease surfaced in the Australian pig population during the first half of the 1980s and ten years later was regarded as one of the most costly and devastating diseases affecting the Australian pig industry.

Task 8.11 Topic sentence analysis

What information would you expect to find in the paragraph introduced by each of the following sentences? What do you think was the focus at the end of the previous paragraph?

1 Propagule pressure is widely recognized as an important factor that influences invasion success (MacDonald et al. 1989; Simberloff 1989; Williamson 1996; Lonsdale 1999; Cassey et al. 2005).
2 Two classes of putative Fe(II)-transport proteins (Irt/Zip and Dmt/Nramp) have been identified in plants (Belouchi et al., 1997; Curie et al., 2000; Eide et al., 1996; Thomine et al., 2000).

Check the paragraphs in the PEAs by Britton-Simmons and Abbott and Kaiser et al. (Chapters 18 and 19) to find out if your predictions are correct, and see also our comments in the Answer pages.

Look at an article you have not read before and read the first sentences of each of the paragraphs in the Introduction. Can you predict the content of the paragraphs? N.B. The first sentence is very often but not always the topic sentence of the paragraph.

Do you agree that Sentence 4 is more *general* than the other sentences? In that case, the paragraph could be improved by moving Sentence 4 to the beginning of the paragraph, as below. Some slight changes of wording have also been made to improve the sense.

> Pleuropneumonia (APP) surfaced in the Australian pig population during the first half of the 1980s and ten years later was regarded as one of the most costly and devastating diseases affecting the Australian pig industry. It can present as a dramatic clinical disease or as a chronic, production limiting disease in pig herds. A sudden increase in the number of sick and coughing pigs and a sharp rise in mortalities among grower/finisher pigs may herald an outbreak of APP in a herd. On the other hand, signs may be limited to a drop in growth rate and an increase in grade two pleurisy lesions in slaughter pigs.

Strategy 3: Put old (or given) information before new information

To understand the basis of this recommendation, consider first the two short paragraphs below. Both contain exactly the same information, but in a different order: decide whether one version is easier to understand than the other.

> **Version A** [1]Clay particles have surface areas which are many orders of magnitude greater than silt or sand sized particles. [2]The ability of soils to shrink when dried is controlled by the interactions of these clay surfaces with water and exchangeable cations.

> **Version B** [1]Clay particles have surface areas which are many orders of magnitude greater than silt or sand sized particles. [2]The interactions of these clay surfaces with water and exchangeable cations control the ability of soils to shrink when dried.

Readers usually agree that version B is easier to follow. The following section seeks to explain why this should be so. When readers begin to read sentence 2 of either version of the paragraph, they already know all the information that is included in sentence 1; therefore all the sentence 1 information can be described as old or given information in this context. In version A, it is not till the second half of sentence 2 that readers encounter a reference to this old information again (clay surfaces). All the information at the beginning of sentence 2 is new information, and so the sentence does not follow the recommendation to put old information before new information. This structuring contributes to making the passage difficult to follow. In version B, the information order has been changed to put the old information at the beginning of sentence 2 and the new information at the end.

Task 8.12 Old information before new information

Which sentence needs changing to follow the guideline given above?

Pleuropneumonia (APP) surfaced in the Australian pig population during the first half of the 1980s and ten years later was regarded as one of the most costly and devastating diseases affecting the Australian pig industry. It can present as a dramatic clinical disease or as a chronic, production limiting disease in pig herds. A sudden increase in the number of sick and coughing pigs and a sharp rise in mortalities among grower/finisher pigs may herald an outbreak of APP in a herd. On the other hand, signs may be limited to a drop in growth rate and an increase in grade two pleurisy lesions in slaughter pigs.

Check your answer in the Answer pages.

Strategy 4: Make a link between sentences within the first seven to nine words

Another way to describe the difference between versions A and B under Strategy 3 relates to how many words the reader has to read in the next sentence (sentence 2 in each version) before encountering a link with what is already known (the old information). In version A, the reader has to read 15 words before finding the first link, which is the word *clay*. In version B, however, the first link word comes as word five of sentence 2. Making this link within the first seven to nine words of sentences enhances the readability of the writing: that is, the ease with which readers will process the information presented. Sentence 3 in Task 8.12 works better when it is re-written as follows.

> An outbreak of APP in a herd may be heralded by a sudden increase in the number of sick and coughing pigs and a sharp rise in mortalities among grower/finisher pigs.

In this version, the fourth word (APP) provides the old information, and old information precedes new information. The method used to change the information order in the sentence was to change an active voice verb, may herald, to a passive voice verb, may be heralded. This method is often useful to improve flow within paragraphs. In our opinion, promoting flow in this way is a more important consideration that avoiding the passive voice at all costs, as is sometimes recommended in writing manuals.

Strategy 5: Try to include the verb and its subject in the first seven to nine words of a sentence

Read the following two sentences and consider how easy they are to follow.

> [1]The definition of seed quality is very broad and encompasses different components for different people. [2]The quality and quantity of flour protein, dough mixing

requirements and tolerance, dough handling properties and loaf volume potential are quality parameters of wheat seed for bread bakers.

Sentence 2 is not easy to follow because readers have to read a very long subject of 19 words before they arrive at the verb *are*. Sentences with very long subjects and short verbs at the end are often called top-heavy sentences. In both the edited versions below, sentence 2 has been changed so that the verb and its subject fit within the first seven to nine words, and the list of items (which makes up the new information in the sentence) comes at the end.

Edited version A [1]The definition of seed quality is very broad and encompasses different components for different people. [2]Quality parameters of wheat seed for bread bakers are the quality and quantity of flour protein, dough mixing requirements and tolerance, dough handling properties and loaf volume potential.

Edited version B [1]The definition of seed quality is very broad and encompasses different components for different people. [2]For bread bakers, quality parameters of wheat seed are the quality and quantity of flour protein, dough mixing requirements and tolerance, dough handling properties and loaf volume potential.

As a general rule, if you want to write a list, it should come at the end of its sentence.

Task 8.13 Revising top-heavy sentences

Change these top-heavy sentences so that each has a verb and its subject within the first seven to nine words.

1 In this project the *Rhizoctonia* populations of two field soils in the Adelaide Plains region of South Australia were characterised.
2 A balance between deep and shallow rooting plants, heavy and light feeders, nitrogen fixers and consumers and an undisturbed phase is needed to achieve maximum benefit through rotation.

Compare your answers with the suggested improvements in the Answer pages.

Task 8.14 Revising your own Introduction for flow

If you are writing a draft Introduction as you proceed through this book, take time now to revise it using the strategies discussed in Chapter 8.

CHAPTER 9

The Discussion section

9.1 Important structural issues

There are several important issues to think about as you begin to draft your Discussion section.

Structure of the Discussion

- Does the journal you are targeting allow the option of a combined Results/Discussion section, followed by a separate Conclusion? Would this arrangement suit your story?
- Does the journal permit a Conclusion where the Discussion is relatively long? Would your paper benefit from one?
- Does the journal publish Discussion sections which include subheadings? Would this option help you signal your main messages to the reader?

Relating the Discussion closely to the paper's title

- As you decide on the key elements of the paper's story that will be emphasized in the Discussion, consider redrafting the title to reflect them more clearly.

Relating the Discussion closely to the Introduction

- Remember that you need to ensure that your Discussion connects clearly with the issues you raised in your Introduction, especially the country where you began (see section 8.2), the evidence leading up to your Stage 3 gap or research niche, and your statement of purpose or main activity. When the first draft of the Discussion is ready, go back to the Introduction and check for a close fit. If necessary, redraft the Introduction to make sure the issues of importance in the Discussion appear there also.
- However, it is not necessary to include in the Introduction all the literature that will be referred to in the Discussion. It is important not to repeat information unnecessarily in the two sections.

Writing Scientific Research Articles: Strategy and Steps, 1st edition. By M. Cargill and P. O'Connor. Published 2009 by Blackwell Publishing, ISBN 978-1-4051-8619-3 (pb) and 978-1-4051-9335-1 (hb)

> ### Task 9.1 Structures of Discussion sections
>
> Check the Discussion section of your selected PEA.
>
> - Does it include subheadings?
> - Is it followed by a separate section headed Conclusion(s)?
>
> Now answer the same questions about your SA. Discuss your findings with a colleague or teacher if appropriate. Why do you think the author chose the arrangement they did? Do you think the Discussion could have been improved by using a different arrangement?

9.2 Information elements to highlight the key messages

The types of information commonly included in Discussion sections are given below: this list can form a checklist for you as you write. You may not have something to say under every point in the list for every result you discuss, but it is worthwhile thinking about each element in turn as you draft the section.

1 A reference to the main purpose or hypothesis of the study, or a summary of the main activity of the study.
2 A restatement or review of the most important findings, generally in order of their significance, including

  i whether they support the original hypothesis, or how they contribute to the main activity of the study, to answering the research questions, or to meeting the research objectives; and
  ii whether they agree with the findings of other researchers.

3 Explanations for the findings, supported by references to relevant literature, and/or speculations about the findings, also supported by literature citation.
4 Limitations of the study that restrict the extent to which the findings can be generalized beyond the study conditions.
5 Implications of the study (generalizations from the results: what the results mean in the context of the broader field).
6 Recommendations for future research and/or practical applications.
  (After Weissberg and Buker 1990).

The elements numbered 2–5 are often repeated for each group of results that is discussed.

> ### Task 9.2 Information elements in the Discussion section
>
> Select the part of this task, 1 or 2, that relates to your selected PEA.
>
> 1 From Kaiser et al. (2003) (provided in Chapter 18), read the second subsection of the Discussion, under the heading *Specificity of GmDmt1;1*. For each sentence, and based on the checklist given above, identify the information element(s) that are presented.
>
> *(Continued)*

Task 9.2 (*Continued*)

2 From Britton-Simmons and Abbott (2008) (provided in Chapter 19), read the first paragraph of the Discussion. For each sentence, and based on the checklist given above, identify the information element(s) that are presented.

Check your answers in the Answer pages.

Task 9.3 Analyzing a Discussion section

Select one or more paragraphs from the Discussion section of your SA to use for a similar analysis to the one you performed for Task 9.2.

- For each sentence, identify the information element(s) that are presented.
- Can you identify any strategies the authors have used to clarify the key messages of their Discussion section (subheadings, topic sentences)?
- Is there a close link between the key or "take-home" messages and the paper title?

Discuss your findings with a colleague or teacher if appropriate.

Task 9.4 Drafting your own Discussion section

Begin to draft the Discussion section of your own paper, if appropriate, using the checklist in Section 9.2 to ensure you include all the relevant information elements.

When drafting this section, it can be useful to think about the main points you want your reader to understand from the Discussion, and consider using subheadings or topic sentences to highlight where the discussion focuses on each of these points.

9.3 Negotiating the strength of claims

For the last four information elements mentioned above, authors need to pay particular attention to the *verbs* they use to comment on their results. The verbs carry much of the meaning about *attitude to findings* and *strength of claim*.

In sentences using *that*, authors have two opportunities to show how strong they want their claim to be:

- in the choice of vocabulary and tense in the main verb;
- in the choice of verb tense in the *that* clause.

Let us look at some example sentences from the PEAs (Tables 9.1–9.4). The verb phrases of interest are underlined in the tabular presentations of Examples 1–4 below.

In Example 1 (Table 9.1), the main verb is in the present tense (indicating that it is "always true", a very strong statement) and the meaning of the verb itself

Table 9.1 Example 1 of language choices in a Discussion sentence.

Subject of main verb	Main verb	"That" plus subject of "that" clause	Verb from "that" clause	Rest of sentence
Our experimental results	demonstrate	that space- and propagule-limitation both	regulate	*S. muticum* recruitment.

Table 9.2 Example 2 of language choices in a Discussion sentence.

Subject of main verb	Main verb	"That" plus subject of "that" clause	Verb from "that" clause	Rest of sentence
These results	indicate	that *S. muticum* recruitment under natural field conditions	will be determined	by the interaction between disturbance and propagule input.

Table 9.3 Example 3 of language choices in a Discussion sentence.

Subject of main verb	Main verb	"That" plus subject of "that" clause	Verb from "that" clause	Rest of sentence
…it	appears	that GmDmt1;1	has	the capacity to function *in vivo* as either an uptake or an efflux mechanism in symbiosomes.

Table 9.4 Example 4 of language choices in a Discussion sentence.

Subject of main verb	Main verb	"That" plus subject of "that" clause	Verb from "that" clause	Rest of sentence
The presence of an IRE motif	suggests	that GmDmt1;1 mRNA	may be stabilized	by the binding of IRPs in soybean nodules when free iron levels are low.

(*demonstrate*) is also strong; the verb in the *that* clause is also in the present tense. Together, these choices indicate that the authors are very confident of the claim they make in this sentence. That is, they think that the data they have presented in the article are strong enough to justify making the strongest possible statement about what the results mean.

Example 2 (Table 9.2) is of similar strength to Example 1: *indicate* is similar in strength of certainty to *demonstrate*, and present tense is used in the main clause;

the verb in the *that* clause is in the future tense, indicating a strong prediction of outcome.

In Example 3 (Table 9.3) a much weaker verb is used in the main clause: *appears* (which is only ever used with the subject *it* in this kind of sentence). The verb in the *that* clause is in the present tense, reflecting the strength of the evidence the authors have presented earlier in the paragraph.

In Example 4 (Table 9.4), the main clause verb *suggests* is again weak in terms of its level of certainty; in addition, the verb in the *that* clause has been made less definite by the use of the modal verb *may*. Thus Example 4 makes the weakest claim of any of the sentences we have considered here. This is not a bad thing at all: the important thing for authors is that they match the strength of their sentences (using the vocabulary and tense options discussed above) with the strength of the data and arguments they have presented in the Results and Discussion sections of the paper. This is a key feature that is checked by referees during review of a manuscript, and by thesis examiners as well.

Task 9.5 Negotiating strength of claims with verbs

Complete the schematic in Table 9.5 by listing alternative choices for the underlined words, writing them in increasing order of strength down the page. The strongest alternatives have been completed as an example.
  Check your answers with our suggestions in the Answer pages.

Table 9.5 Task 9.5: Negotiating strength of claims with verbs, an exercise in ranking possible verb forms in a Discussion sentence in order of strength of claim.

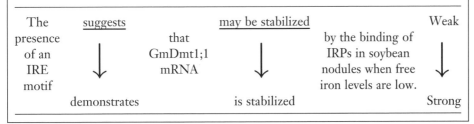

An alternative construction without a *that* clause is also found in science writing. Look at the example below, taken from the PEA by Britton-Simmons and Abbott (2008).

> Previous studies have demonstrated a positive relationship between propagule pressure and the establishment success of non-native species.

In this construction, the object of the verb is a noun phrase, here "a positive relationship between propagule pressure and the establishment success of non-native species". It is interesting to note that when this construction is used, the author does not need to make a decision about what tense to use in the *that* clause.

Task 9.6 Analyzing and practicing strength of claim

Reread the Discussion section of your selected PEA and find sentences that use both these patterns in the Discussion or Conclusion sections. Identify the verbs that carry the strength-of-claim messages, and discuss your findings with a colleague or teacher, if appropriate.

Then consider your own results and begin to draft sentences to comment on them in your Discussion section, paying particular attention to matching the strength of your claim in your sentences to the strength of your data and arguments.

CHAPTER 10

The title

The title you finally select for your manuscript forms an important part of your communication with your readers, both with the editor and referees who will evaluate the paper, and with the members of your discipline community whom you want to read the paper after its publication. From the referee criteria we considered in Chapter 3 we know it is important that the title clearly indicates the content of the paper, but there are various ways in which that can be achieved. In this chapter we look at advice about attracting the attention of your target readers effectively.

10.1 Strategy 1: Provide as much relevant information as possible, but be concise

The purpose of a title is to attract busy readers in your particular target audience, so that they will want to access and read the whole document. The more revealing your title is, the more easily your potential readers can judge how relevant your paper is to their interests. To exemplify the importance of this issue, we quote from relevant Author Guidelines: the *Journal of Ecology* asks for "a concise and informative title (as short as possible)" (www.blackwellpublishing.com/submit. asp?ref=0022-0477&site=1); the *New Phytologist* stipulates a concise and informative title (for research papers, ideally stating the key finding or framing a question; www.blackwellpublishing.com/submit.asp?ref=0028-646X&site=1). We will return to this question of the most effective grammatical form for titles later.

10.2 Strategy 2: Use keywords prominently

It is important to decide which words (keywords) will capture the attention of readers likely to be interested in your paper and to place them near the front of your title. This practice also helps ensure that your title is picked up efficiently by the literature-scanning services, which use a keywords system to identify papers of interest to particular audiences. Wherever possible it is a good idea to place the

Writing Scientific Research Articles: Strategy and Steps, 1st edition. By M. Cargill and P. O'Connor. Published 2009 by Blackwell Publishing, ISBN 978-1-4051-8619-3 (pb) and 978-1-4051-9335-1 (hb)

most important word(s) in your title in the position of power: the beginning. For example:

✗ Effects of added calcium on salinity tolerance of tomato
✓ Calcium addition improves salinity tolerance of tomato

One effective way to ensure your keyword(s) are at the front of your title is to use a colon (:) or a dash (–) to separate the first, keyword-containing part of the title from a second, explanatory section. Effective examples include the following (taken from the reference lists of the PEAs):

✓ Disturbance, invasion, and reinvasion: managing the weed-shaped hole in disturbed ecosystems
✓ Native weeds and exotic plants: relationships to disturbance in mixed-grass prairie
✓ Methylamine/ammonium uptake systems in *Saccharomyces cerevisiae*: multiplicity and regulation
✓ Resistance to infection with intra-cellular parasites – identification of a candidate gene

10.3 Strategy 3: Choose strategically: noun phrase, statement, or question?

The traditional way to write titles and headings is as a noun phrase: a number of words clustered around one important "head" noun. Below are some examples of this kind of title, with the head nouns shown in bold.

- **Diversity and invasibility** of southern Appalachian plant communities
- Food expenditure **patterns** in urban and rural Indonesia
- **Systems** of weed control in peanuts
- Iron **uptake** by symbiosomes from soybean root nodules
- **Evidence** of involvement of proteinaceous toxins from *Pyrenophora teres* in net blotch of barley

Several of these titles are very effective: brief, informative, and with keywords placed near the front. However, this style of title writing is not always the best for meeting the two guidelines discussed under Strategies 1 and 2 above. Look again at the last title in the list, "Evidence of involvement of proteinaceous toxins from *Pyrenophora teres* in net blotch of barley". This title leaves us with an unanswered question: what kind of involvement? Additionally, the first four words are very general in meaning, giving no enticement to the reader to continue reading. Rewriting this title as a statement could overcome these difficulties, and was in fact recommended by a referee when this paper was under review. (A statement is a sentence with a subject and a verb, and its advantage in this context is that it can give more explicit information about the results of the study.)

✗ Evidence of involvement of proteinaceous toxins from *Pyrenophora teres* in net blotch of barley
✓ Proteinaceous metabolites from *Pyrenophora teres* contribute to symptom development of barley net blotch (Sarpeleh et al. 2007)

Statement titles are only suitable for papers that address a specific question and present a non-complex answer. In these conditions, the sentence form is a good option to replace titles that begin with vague terms such as "The effects of...". For example:

   ✗   Effects of added calcium on salinity tolerance of tomato
   ✓   Calcium addition improves salinity tolerance of tomato

When there is no simple answer to be presented, it can be effective to write a title as a question, e.g.:

   ✓   Which insect introductions succeed and which fail?

As with all sections of your manuscript, check whether the journal has specific conventions or recommendations about the form of titles before you decide which form to use. In our own experience, it can be useful to develop a list of possible titles as you draft your manuscripts, and choose the most effective one for the target audience and the paper's key message right at the end of the writing process.

10.4 Strategy 4: Avoid ambiguity in noun phrases

If writers place a string of nouns and adjectives together, to form a title which packs a lot of meaning into a few words, they can sometimes cause problems of ambiguity: more than one possible meaning. This is particularly the case when nouns are used as adjectives, i.e. placed in front of the head word of the noun phrase. To investigate why this is so, let's consider some examples.

The noun phrase germination conditions has only one possible meaning: conditions for germination, and thus it can be used without risk of ambiguity. Similarly, application rate can only mean the rate of application. However, enzymatic activity suppression could mean either suppression *of* enzymatic activity or suppression *by* enzymatic activity and is therefore ambiguous. A general guideline is to restrict these noun phrases to a maximum of three words, and this many only if there is no risk of misunderstanding. If they grow longer, rewrite them by inserting the prepositions that clarify the meaning (e.g. of, by, for). For example:

   ✗   soybean seedling growth suppression
   ✓   suppression of soybean seedling growth

N.B. When nouns are used as adjectives in extended noun phrases, they are always used in the singular. Useful examples to help you remember this are as follows.

food for dogs	→	dog food
disturbance by herbivores	→	herbivore disturbance
nodules on soybean roots	→	soybean root nodules

Task 10.1 Analyzing article titles

Complete Table 10.1 and discuss your findings with colleagues or teachers if appropriate. Compare your answers with our suggestions in the Answer pages. Now, spend a little time deciding if there are any improvements you can make to the title you have drafted for your OA.

Table 10.1 Task 10.1: Analyzing article titles, an exercise in analyzing the structure and communicative effectiveness of selected article titles.

Question	Kaiser et al. (2003)	Britton-Simmons and Abbott (2008)	Your selected article
Is the title a noun phrase, a sentence, or a question?			
How many words are used in the title?			
What is the first idea in the title?			
Why do you think this idea has been placed first?			

CHAPTER 11

The Abstract

11.1 Why Abstracts are so important

- For busy readers the Abstract, sometimes called the Summary, may be the only part of the paper they read, unless it succeeds in convincing them to take the time to read the whole paper!
- For readers in developing countries with limited access to the literature, the Abstract may be the only information on your work that is available to them.
- Abstracting services may use the text of the title plus the Abstract and keywords for their searchable databases.

11.2 Selecting additional keywords

Consult other similar papers in your field to see which additional keywords they use beyond the ones already included in the title. The idea is to select from the list used by the relevant indexing services. At this stage, think again about your audience and their interests, and try to predict what keywords they might use to search under.

11.3 Abstracts: typical information elements

Some journals provide a list of questions or headings for authors to respond to in writing their abstracts, and others do not. All provide a maximum number of words that an abstract (or summary) may contain (e.g. 250 for *The Plant Journal* and 350 for the *Journal of Ecology*, as of March 2008). Based on analyses of many abstracts in science and technology fields, the following information elements can be proposed as constituting a full abstract or summary (Weissberg and Buker 1990).

Some background information	B
The principal activity (or purpose) of the study and its scope	P
Some information about the methods used in the study	M
The most important results of the study	R
A statement of conclusion or recommendation	C

Writing Scientific Research Articles: Strategy and Steps, 1st edition. By M. Cargill and P. O'Connor. Published 2009 by Blackwell Publishing, ISBN 978-1-4051-8619-3 (pb) and 978-1-4051-9335-1 (hb)

This list is often compressed to the following components.

Principal activity/purpose and method of the study	P + M
Results	R
Conclusion (and recommendations)	C

Task 11.1 Analyzing Summaries

Read the Summaries of both the PEAs and identify which of the information elements listed above are present, and in which sentence(s). (Even if you are not completely familiar with the science being presented in both papers, these sections are short enough that you should be able to complete this task without difficulty, and there are important things to learn from doing so.)

Compare your answers with our suggestions in the Answer pages.

N.B. The *Journal of Ecology*, which published the Britton-Simmons and Abbott paper, provides the following guidelines for the writing of the Summary:

> *Summary* (called the Abstract on the web submission site). **This must not exceed 350 words** and should list the main results and conclusions, using simple, factual, numbered statements. The final point of your Summary must be headed 'Synthesis', and must emphasize the key findings of the work and its general significance, indicating clearly how this study has advanced ecological understanding. This policy is intended to maximize the impact of your paper, by making it of as wide interest as possible. This final point should therefore explain the importance of your paper in a way that is accessible to non-specialists. We emphasize that the Journal is more likely to accept manuscripts that address important and topical questions and hypotheses, and deliver generic rather than specific messages. (www.blackwellpublishing.com/submit.asp?ref=0022-0477&site=1, retrieved 28 March 2008)

The final sentence of this advice is particularly relevant to us in our analysis of this paper, as it provides a rationale for what has been emphasised in the strategically important parts of the paper – the title, the summary, the end of the introduction and the discussion. This fact underlines how very important it is to seek out, read carefully and respond effectively to the *Author Guidelines* (or equivalent) for the journal to which you will submit your manuscript.

Task 11.2 Analyzing your SA Abstract or Summary

Repeat Task 11.1 for your SA, and discuss your findings with a colleague or teacher, if appropriate.

Task 11.3 Drafting your own Abstract or Summary

Now write or revise your own Abstract or Summary, if appropriate. One way to begin is to write sentences for all of the information elements given above and then combine them into a first draft of your Abstract. Then check the number of words you have used against the requirement of the journal you are targeting. If necessary, shorten your draft, using techniques such as those you have observed in the Abstracts/Summaries you have analyzed.

Getting your manuscript published

CHAPTER 12

Considerations when selecting a target journal

Choosing the right journal for your manuscript will influence the chance of getting published easily and quickly. You should be thinking about the journal you want to publish in from the beginning of your research and should have made a choice by the time you begin to write the Introduction and Discussion sections of your paper.

The choice of journal determines the size of the audience who can access and use your work and the professional prestige and rewards which may flow from the publication. The right journal for you is the journal which optimizes the speed and ease of publication, the professional prestige you accrue, and the access for your desired audience. These factors are interwoven and it can be helpful to develop a publication plan to maximize your publication success. As discussed in Chapter 1, one of the first considerations is whether the journal peer reviews the articles that it publishes. The peer-review process is important for establishing the quality of your work, and you should seek peer-reviewed journals to publish in if you wish to develop a research profile. Of course, the journal of your choice may not choose to accept your article, and you are advised to have a list of preferred journals to turn to if you are rejected from your first choice. Here we set out some issues to consider when choosing a journal for your manuscript.

12.1 The scope and aims of the journal

The journals that are most often cited in the Introduction and Discussion sections of your manuscript will be most likely to accept work in your field. Examine some of the key articles you refer to in Stages 2 and 3 in your Introduction, and check which journals are cited in Stages 2 and 3 of the Introductions of these articles. By following back through the literature you should be able to develop a mind-map of the journals in the field of your research. Check the websites or issues of these journals to identify those with scope and aims most appropriate for your manuscript.

12.2 The audience for the journal

The audience for a journal is largely determined by the scope and aims of the journal, the journal's reputation and history of publishing in the field, and the

Writing Scientific Research Articles: Strategy and Steps, 1st edition. By M. Cargill and P. O'Connor. Published 2009 by Blackwell Publishing, ISBN 978-1-4051-8619-3 (pb) and 978-1-4051-9335-1 (hb)

accessibility of the journal to researchers (e.g. is it expensive, does it have Open Access options for authors, is it published by a small publisher with limited distribution?). Internet access to journal titles, abstracts, and homepages has allowed many more journals to be accessible to a wider audience. However, some users may not wish to pay for access to a paper, and so journals that are widely bought by institutions will have a wider audience for practical purposes. New journals may also take time to develop an audience. Check the journal website and publisher to see whether a journal you are considering is widely distributed.

12.3 Journal impact

There is no easy way to assess the quality of a journal or the contribution of a journal to a research discipline over time. A number of indices have been developed to provide information on the relative speed and volume of citation to journals, and these indices can give some guidance about the relative popularity and usage of a journal. The most commonly used measure of journal impact is the *Journal Impact Factor*.

The Journal Impact Factor for a given year is the average number of times articles published in the journal in the two previous years have been cited in that year. This index provides a measure of the average recent use of articles in a given journal. It is calculated using the following formula.

$$\text{Journal Impact Factor (Year}_x) = \frac{\text{Cites to recent articles (Year}_{x-1} + \text{Year}_{x-2})}{\text{Number of recent articles (Year}_{x-1} + \text{Year}_{x-2})}$$

Other measures of the influence of a journal on its field of research are

- *Journal Immediacy Index*, calculated as the number of citations to articles in the year with respect to the number of articles published in that year, giving a measure of how rapidly the average article in a given journal is used;
- *Journal Cited Half-Life*, calculated as the number of publication years from the current year that account for 50% of citations received by the journal, giving a measure of the longevity of use of the average article in a given journal.

12.4 Using indices of journal quality

Statistics on citation number as a measure of journal quality should be used with an awareness of the purpose for which the statistics are gathered and the limitations of these indices. The indices described above all measure the rate or volume of citation of the *average article* in a journal. They are measures of the journal and not the individual articles. The number of citations for your article can also be calculated and may be higher or lower than the average for the journal. Getting your articles read and cited (or used) is about reaching the right audience. Sometimes the right audience may not be the readership of the journal with the highest impact factor.

Other things to consider when assessing indices for ranking journals are these.

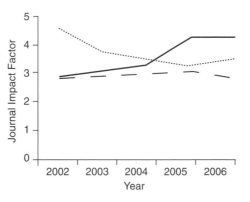

Fig. 12.1 Trend of Journal Impact Factor for three different journals in the plant sciences (*source*: ISI Web of Knowledge, Journal Citation Reports 2008).

- Comparing journals from different fields of research may not be meaningful (e.g. mathematics researchers cite very few journals, whereas papers in molecular biology journals cite dozens).
- The calculation of some indices is prone to inflate the relative contribution of journals which include sections for discussion and review (rather than original research).
- Citation-matching procedures are strongly affected by sloppy referencing, editorial characteristics of journals, some referencing conventions, language problems, author-identification problems, and unfamiliarity with names from some countries.
- Published indices are calculated from a selected list of journals. This list largely excludes journals published in non-English-speaking countries, and may not include new journals still establishing their reputation.
- Journal ranking based on indices can change over time. Figure 12.1 shows the Journal Impact Factors for three popular journals in the plant sciences over a 3-year period. The impact factor for one journal increased, one decreased, and one remained relatively stable. However, articles in each of the three journals will continue to be cited on their individual merit.

12.5 Time to publication

Journals want to publish submissions quickly to ensure they attract authors who are doing innovative and new work. You may also want to publish your research quickly to ensure that others do not publish similar work before you, and to increase your publication and citation record for promotions and grants. If time to publication is important to you, you should check journal websites or recent issues to see whether they report the average time to publication. Journals which publish an online version of the paper before the print version will usually have a faster time to publication.

12.6 Page charges or Open Access costs

Some journals charge fees for publishing manuscripts. Fees may be based on a fixed cost or on the number of pages, or they may be charged for publishing colour illustrations or for reprints. Check whether the journal charges for any part of the publishing process before you submit your manuscript. You may also want your research to be accessible to a wide range of readers who do not have access to libraries or other subscriptions to journals in your field. Many journals now offer to provide Open Access to your paper (i.e. to make it accessible for free download without subscription to the journal) if you pay an upfront fee. Check whether the journal of your choice offers this service if you want (or are required by your institution) to pay for Open Access.

Task 12.1 Analyzing potential target journals

To optimize the outcomes from publishing your manuscript, we recommend that you develop a publishing strategy. Part of the publication strategy is to select your preferred journal to submit the manuscript to. In order to make this choice, first select the three or four preferred journals in your field that you think would accept your manuscript. Then answer the following questions for each one and record the answers in Table 12.1.

1 Has the journal published similar work with a similar level of novelty to yours in the last 3 years? Record a yes or no (if "no", think carefully before submitting your manuscript to this journal).
2 Does the journal's scope and the content of recent articles match the main components of your manuscript, i.e. subject, methods, results? (Write down the main type of papers, e.g. plant physiology: non-molecular studies).
3 What is the measure of relative journal quality/impact which is most important to you and your field of research? Record the score or measure for each journal (e.g. Journal Impact Factor or Journal Cited Half-Life).
4 What is the journal's time to publication? (This may be on the journal's website or recorded for each article in the journal.) Record the time or a score for fast or slow (e.g. less than 3 months from acceptance = fast; more than 1 year = slow).
5 Does the journal have page charges or provide Open Access if you want it (and can you pay if payment is required)?

Examine the journal scores you have recorded in Table 12.1 and rank the journals in order of overall preference, taking all criteria into consideration.

Table 12.1 Rating preferred journals in terms of key criteria for maximizing your publication success.

Journal name	Recent publication of similar work and novelty	Match of scope and recent content to your work	Journal quality/impact	Time to publication	Page charges or Open Access costs
1					
2					
3					
4					

CHAPTER 13

Submitting a manuscript

Submitting your manuscript to a journal is like entering any competition where success is determined by a group of judges using a defined set of selection criteria. You can optimize your publication success by understanding and meeting the selection criteria of the journal. Many of the selection criteria related to manuscript preparation will be listed by the journal on their website or in printed issues of the journal (e.g. Instructions to Contributors and journal scope or aim). Other criteria relate to how a manuscript conforms to the standard of the journal and can only be understood by reading and thinking about the journal and by understanding the editing and review processes. Here we describe the editing and reviewing of journal articles and document the main selection criteria used by editors and referees. This information will help you to adopt practices that will help you develop your publishing strategy and navigate the publishing process, leading to publication success.

13.1 Five practices of successful authors

Success as a scientist is largely measured by the quality and quantity of research output and the impact of that research on other research or practice. Publishing scientific articles is a necessary part of success as a scientist. Successful authors adopt five practices to optimize their publication outcomes. They

1 review manuscripts for colleagues and journals and develop a strong framework for research writing and manuscript critique;
2 plan their research and writing to meet the quality assurance criteria that referees and editors will impose;
3 carefully select the journal they will submit to and prepare the manuscript content and style to maximize their chances of acceptance;
4 use structured review processes and pre-reviews from colleagues to improve the manuscript before submitting it to a journal; and
5 use journal referee reports to improve the manuscript and demonstrate to the journal editor how improvements have been made.

13.2 Understanding the peer-review process

A scientific research article does not produce truth or certainty but documents the observations/measurements, analysis, and interpretation of the authors in the

Writing Scientific Research Articles: Strategy and Steps, 1st edition. By M. Cargill and P. O'Connor. Published 2009 by Blackwell Publishing, ISBN 978-1-4051-8619-3 (pb) and 978-1-4051-9335-1 (hb)

context of previous research. The veracity of findings from a scientific study will be confirmed by subsequent research or application, and may be qualified or amended over time. The peer-review process assists the scientific community in assuring the quality of research before it is published and before it can be examined and used by a wider audience. Peer reviewing is part of the process of turning information into knowledge. The correspondence between the author, reviewer, and editor is part of a collective sense-making process used to test that new information is worth knowing and acting upon. The system of peer review is not perfect, but it does make a number of critical contributions to the standard of scientific research publications. Specifically, peer review

- confirms that the hypotheses have been tested appropriately and that results reported reflect the materials, methods, and analysis tools used;
- confirms that the strength of claims about the results and the implications of the study are appropriate;
- assists journals to decide whether the focus, novelty, and importance of the research are appropriate for the standard of the journal;
- checks that the presentation and style of the content conforms to accepted conventions for production and reader convenience; and
- advises the authors and the journal editor about how (and often where) the manuscript could be improved.

Referees are important to the journal editor because they take a critical role in determining the quality of manuscripts, and in most cases they do this as a professional contribution and without payment. Referees are important to the author because they bring a critical eye to the content and writing, and highlight how the story can be clarified or more suitably presented. Peer review provides the opportunity to have your ideas, theories, methods, results, analysis, and interpretation considered and commented on by a professional colleague. Responding to the comments of a peer reviewer should be seen as part of the process of testing and legitimizing your research results and their meaning.

The best way to develop your understanding of the peer-review process is to carry out peer review yourself. You may be asked to review for journals if you are publishing your own work. If you are not publishing yet, you can offer to review the work of your colleagues or form a journal club and examine the work of already published authors (see Task 13.2, below). (See Chapter 16 for additional ideas on developing refereeing skills.)

13.3 Understanding the editor's role

The editor is responsible for maintaining the reputation and competitiveness of the journal. Editors use referees to assist them in selecting manuscripts and improving them for publication. The editor will read the manuscript and make the initial decision as to whether it will be sent to reviewers. The editor will usually reject a manuscript without review only if the manuscript is outside the scope or aims of the journal, if the language or structure of the manuscript is poor, or if there are clear or obvious flaws in the science (see Table 14.1 for a guide to dealing with rejection). A well-prepared manuscript reporting science appropriate to the journal is unlikely to be rejected without review. You can use the contributor's covering letter to assist the editor in deciding that your manuscript is ready for review.

The covering letter you send to the editor with your manuscript (or upload in the appropriate box on the journal's submission website) is an important opportunity to sell your paper. The letter is an opportunity to demonstrate that you appreciate the role of the editor and that you have done everything you can to prepare the manuscript to meet the journal's requirements. You can use the covering letter to

- express your belief that the paper is within the scope of the journal;
- state the title of the manuscript and the names of the authors;
- state that the research and the paper are new and original;
- highlight specific points that reinforce the novelty and significance of the research;
- highlight any points about the manuscript which may raise questions for the editor, e.g. that a long paper is justified or that photographs are necessary to report important findings;
- express hope that the presentation is satisfactory; and
- say that you look forward to the referees' comments.

An example covering letter is provided in Figure 13.1.

Task 13.1 The contributor's letter as sales pitch

Examine the example covering letter in Figure 13.1 and draw a box around the words which sell the manuscript to the editor most strongly. Check your answers with our suggestions in the Answer pages.

Date...........

The Managing Editor
Australian Journal of Botany
Address...................

Dear Dr Brown,

Please find attached the manuscript "Arbuscular mycorrhizal associations of the southern Simpson Desert". This manuscript examines the mycorrhizal status of plants growing on the different soils of the dune-swale systems of the Simpson Desert. There have been few studies of the ecology of the plants in this desert and little is known about how mycorrhizal associations are distributed amongst the desert plants of Australia. We report the arbuscular mycorrhizal status of 47 plant species for the first time. The manuscript has been prepared according to the journal's Instructions for Authors. We believe that this new work is within the scope of your journal and hope that you will consider this manuscript for publication in the *Australian Journal of Botany*.

We await your response and the comments of reviewers.

Yours sincerely,

Fig. 13.1 An example covering letter from a manuscript author.

Editors enlist the help of two or more independent researchers to peer review each manuscript and check the quality, novelty, and significance of the work and the presentation of the manuscript. This work is usually unpaid and is undertaken as part of the professional contribution of researchers to the development of their field of science. Reviewers will

- usually be expert in the general field of the paper (not necessary expert in the exact subject of the paper to be reviewed);
- almost always have published work in the general field themselves (possibly work that has been cited in your paper);
- be busy with their own research, writing, teaching, administration, family, etc.;
- be willing to review manuscripts but have limited time and patience; and
- have their own preferences and biases about scientific research and writing.

The journal may have asked you to nominate potential reviewers, or the editor may have chosen them from a database or using professional networks. You will not know who the reviewers are. In many but not all cases, depending on the

---

Referee's Evaluation Form

General questions   Reviewer number: _____

1. Is the contribution new?   ☐ Yes ☐ No
2. Is the contribution significant?   ☐ Yes ☐ No
3. Is it suitable for publication in the Journal?   ☐ Yes ☐ No
4. Is the organization acceptable?   ☐ Yes ☐ No
5. Do the methods and the treatment of results conform
   to acceptable scientific standards?   ☐ Yes ☐ No
6. Are all conclusions firmly based in the data presented?   ☐ Yes ☐ No
7. Is the length of the paper satisfactory?   ☐ Yes ☐ No
8. Are all illustrations required?   ☐ Yes ☐ No
9. Are all the figures and tables necessary?   ☐ Yes ☐ No
10. Are figure legends and table titles adequate?   ☐ Yes ☐ No
11. Do the title and abstract clearly indicate the content
    of the paper?   ☐ Yes ☐ No
12. Are the references up to date, complete, and the journal
    titles correctly abbreviated?   ☐ Yes ☐ No
13. Is the paper excellent, good, or poor?   ☐ Excellent ☐ Good ☐ Poor

Please use a separate sheet for your comments.

Recommendation
☐   Accept without alteration
☐   Accept after minor revision
☐   Review again after major revision
☐   Reject

Reviewer's signature: _____   Date of review: _____

Fig. 13.2 Example evaluation form showing typical questions to which reviewers or referees are asked to respond.

To: Dr AB Brown,
Editor, *Journal of...*

Re: Manuscript Number...
Title...
Authors...

Dear Dr Brown,

The paper describes..... . This is a topic which would benefit from additional work such as that described in the manuscript. However, a major concern with the paper is the interpretation and referencing of the literature in the Introduction and Discussion. Related to this is a lack of integration with previous work to explain aspects of the Methods. The paper needs re-interpretation after a thorough investigation of the literature. I recommend that the paper in its current form be rejected but believe that it may be suitable for your journal after major revision.

Introduction
The Introduction has incorrectly cited [Brown et al. (1981)] who actually showed that......

Methods
Factors relevant to the choice of Methods are: 1) how old were the cultures that were used? 2) Does the age of the culture material affect the results?

Results
The main claim by the authors that their Results showed that... is not correct. Their statement that the results show... needs correction.

Discussion
Relevant references seem to have been overlooked in both the Introduction and Discussion sections, including...

Other queries and suggestions are pencilled on the manuscript.

Yours sincerely,
CD Smith

Fig. 13.3 A referee's report recommending rejection but noting that the paper would be acceptable with some alterations. Content-specific elements have been deleted.

policy of the journal, the reviewers will know the names of the authors. The reviewers will be asked to read the manuscript and write a report about the quality of the work, note any problems, and recommend any changes that would improve the manuscript. The reviewer will usually be asked to complete an evaluation form about the quality of manuscript as well, and may also be asked to recommend whether the manuscript should be accepted by the journal or accepted after revisions. The reviewer will return their written report and the evaluation form to the editor, sometimes with annotations on the manuscript (although this is now less common with electronic submission and review).

Journals have their own set of instructions for reviewers. These are sometimes available on the journal's website, or a colleague who has reviewed for the journal may be able to show them to you. We have constructed an example referee's evaluation form that includes the main questions to which referees are commonly asked to respond (see Figure 13.2). An example of a written report from a reviewer is shown in Figure 13.3.

Task 13.2 Journal club

Form or join a journal club with between three and 10 colleagues in related fields of research and arrange to meet regularly (e.g. once a month). Choose some recent articles of interest to the group and arrange to discuss one of the articles at each meeting. Each member of the group should use the referee evaluation report questions in Figure 13.2 to develop a referee's report on the article. At the journal club meeting, discuss the strengths and weaknesses of the paper and any improvements that could have been made. (See Chapter 16 for additional ideas on journal clubs.)

13.6 Understanding the editor's role (continued)

The editor receives the reports from the referees and decides what response will be made to the author(s). If referees disagree (especially if there are only two referees) the editor will sometimes send the manuscript to a third referee for an additional opinion. The editor then writes to the corresponding author with the decision that has been made. Responding to these letters from journal editors is a skill in itself, and is the subject of Chapter 14.

How to respond to editors and referees

14.1 Rules of thumb

Critical comments about our research or writing can be difficult to accept and respond to. We recommend the following rules of thumb as a framework for responding to referee and editor reports on your manuscript.

Rule 1 It is rare that the referee/editor is completely right and the author completely wrong, or that the author is completely right and the referee completely wrong.

Rule 2 When responding to a referee, the object is to accommodate the referee by addressing their comments without compromising the message (story) of the paper.

Rule 3 Always show the editor that you are doing everything you can to improve the manuscript.

Rule 4 Rejection and criticism do not automatically mean that the science is not good or that the paper is not well written: consider other journals, including additional work, or rewriting some or all of the paper.

14.2 How to deal with manuscript rejection

If your manuscript is rejected it is important to determine the reasons why. The reasons for rejection will inform your decision about how to proceed. Every experienced researcher has a story of rejection, and it can be useful to discuss rejection with a senior colleague to help you see that it is a natural and necessary part of the process of legitimizing scientific knowledge. Almost every one of these colleagues will also tell you that all or some of the data from their rejected manuscripts were eventually published. Remember, *everyone* gets rejections. Successful authors are successful at dealing with rejection as well as acceptance. Reasons why your manuscript may have been rejected and recommendations on how to proceed are listed in Table 14.1.

Writing Scientific Research Articles: Strategy and Steps, 1st edition. By M. Cargill and P. O'Connor. Published 2009 by Blackwell Publishing, ISBN 978-1-4051-8619-3 (pb) and 978-1-4051-9335-1 (hb)

Table 14.1 Reasons for manuscript rejection and recommendations for author response.

Reason for rejection	Response option 1	Response option 2
The content of the paper may not fit the scope of the journal (this could mean it is too specialized, focused on the wrong subject area, or not of enough general interest for the journal's readership).	The editor will have made this decision (usually before it has been reviewed) and it is usually necessary to revise the manuscript and submit it to a more appropriate journal (check your list of preferred journals; see Table 12.1).	If the manuscript has been reviewed, use the referee's reports following Rules of thumb 2 and 3 in section 14.1 and submit it to a more appropriate journal (check your list of preferred journals; see Table 12.1).
There are clear and obvious flaws in the science.	The editor may have made this decision before review. Revise the manuscript using the author response guide (Table 14.2) and try to publish the best parts.	
The language or structure of the manuscript is poor and it could not be sent to referees.	The editor may have made this decision before review. Revise the manuscript using the author response guide (Table 14.2) and resubmit or submit the manuscript to another journal.	
High-ranking journals need to reject a high proportion of submitted manuscripts even if the reviews are (mostly) positive.	Examine the editor's letter and determine whether there is any encouragement to resubmit a revised manuscript (e.g. "revise and resubmit…"). If there is, revise and resubmit following Rules of thumb 2 and 3 (section 14.1).	If resubmission is not encouraged, revise the manuscript using the referee's reports following Rules of thumb 2 and 3 (section 14.1) and submit it to the next journal on your list (see Table 12.1).
Referees may not have read or understood the paper thoroughly enough to appreciate it. Recommendations from the referee to the editor may not be clear or may have been misinterpreted. Something may have annoyed the referee: they are unpredictable and can be helpful or (sometimes) unhelpful.	You can appeal to the editor, although this is unlikely to be successful unless a major error of judgement has been made by the referee or editor. It is always wise to make the uncontroversial changes recommended before appealing, resubmitting, or submitting to a different journal.	Revise and resubmit or submit to a new journal. Clarify any issues which have caused problems by revising the text. If resubmitting to the same journal, make note in your letter to the editor of any misunderstanding, any supportive comments from referees, and the improvements you have made to the manuscript.

14.3 How to deal with "conditional acceptance" or "revise and resubmit"

How to respond to editors and referees

Ch 14
How to respond to
editors and referees

Few manuscripts are accepted for publication without some revision. The level of revision varies from minor changes to the language, references, or formatting to major revisions which may require resubmission for fresh reviewing. In fields where journals compete for a share of the new and interesting research in a discipline and/or a share of the subscription market, journal editors aim to accept high-quality manuscripts as quickly as possible and get them into print in a timely manner. When the science is obviously interesting and new but the manuscript requires major work before it is acceptable, the editor may reject the paper but encourage rewriting and resubmission. If the manuscript requires some modification but not major restructuring, additional research, or rewriting, the editor may accept the manuscript on the condition that recommended changes are made and the article returned by a set date. An example of a conditional acceptance letter is provided in Figure 14.1. This conditional acceptance provides you the opportunity to consider and incorporate the comments of reviewers and the editor. However, it is not always easy to understand or address reviewer comments.

N.B. The difference between a conditional acceptance and a revise and resubmit response is usually only visible in the wording the editor uses in their letter to you. Your first task when you receive the response is to decide what the editor means, and this is not always easy. The editor may use indirect language in the interest of being polite and maintaining your good opinion of the journal for future occasions. If you are in any doubt about the meaning, show the letter to a colleague and discuss it. In any case, you will need to communicate with your co-authors about your response while deciding your strategy for the next stage in the process.

There are many ways to deal with reviewers' comments and you will develop your own strategies. Here we outline an approach used by many experienced authors.

- Don't get angry or offended by the comments. The reviewer or editor may have misunderstood something or you may have communicated it poorly. Dealing with reviewers' comments is part of the publishing process and they should not be seen as a personal attack on your credibility as a scientist.
- Read the comments and check the manuscript to make sure you understand what the referee or editor is asking you to do.
- Highlight any comments which are difficult to respond to or are unclear.
- Show the difficult comments to co-authors or colleagues and seek their advice about how to deal with them. If comments are still difficult, unclear, or they annoy you, leave them for a few days (not more than a week) and return to address them when you have had time to absorb them.
- Review the Rules of thumb (section 14.1).
- Make all the small changes which do not require major rewriting and note each change in a letter to the editor.
- Respond to any major comments using the suggested responses in Table 14.2.

Main types of reviewer comments

Every review is different and will present different challenges to which to respond. However, the majority of reviewer comments fall into the seven categories listed overleaf.

From: Dr AB Brown,
Editor, *Journal of...*

Dear Dr Zhu,

I enclose the referees' reports on your paper entitled .... The referees agree that the paper contains much good material. However, they have recommended that it needs considerable revision before it can be published. In particular, I draw to your attention the following comments by the referees.

Referee 1:
- The Methods section does not give sufficient information, particularly about the sampling methods used.
- The results in Tables 1 and 2 are closely related and can be combined into a single table.
- The conclusion that there is a strong positive correlation between the number of organisms and soil salinity needs a stronger statistical basis.
- The results in Figure 3 are very preliminary - this really requires another survey. If this is not possible, the Figure should be deleted.

Referee 2:
- There are inadequacies in the Methods section, as indicated on the typescript.
- The Discussion is not well focused and does not include some important relevant publications, e.g. Jones et al. (2000). '........' in the *Journal of ...*
- The conclusion is interesting but can be greatly strengthened. In particular, the findings are different from those of Walter et al. (1997) in the *Journal of...*, a study done in the USA. The work in your paper is in fact the first study of its kind outside Europe and North America and this should be highlighted.

There are other comments in the enclosed reports, and some corrections have been made to the English on the typescripts. If you can revise the paper along the lines suggested and resubmit by ... then I will consider its acceptability for publication in the Journal without further reference to referees. However, additional refereeing may be necessary.

I look forward to hearing from you.

Yours sincerely,
AB Brown

Fig. 14.1 An adapted example of a conditional acceptance letter from a journal editor. (N.B. This is an unusually short letter.)

1 The aims of the study are not clear.
2 The theoretical premise or "school of thought" on which the work is based is challenged.
3 The experimental design or analysis methods are challenged.
4 You are asked to supply additional data or information that would improve the paper.
5 You are asked to remove information or discussion.
6 The conclusions are considered incorrect, weak, or too strong.
7 The referee has unspecific negative comments, e.g. "poorly designed", "poorly written", "badly organized", "tables are too large", "relevant literature not cited", or "English is poor".

Decide which of these categories each of the difficult comments falls into. If these categories do not cover the comment you have received, decide what

Table 14.2 Author response guide to how and where (in the manuscript and in correspondence with the editor) to deal with referee reports.

Comment type	Author response	Where in manuscript	Section to check in this book
The aims of the study are not clear.	Rewrite the aims to state them clearly. Ensure the aims are consistent with, and linked to, the experimental design. Ensure the Discussion refers back to the aims.	Introduction (Stage 4) Cross-check Introduction (Stage 4) with Methods Cross-check Discussion with Introduction (Stage 4)	Section 8.6 Chapters 7 and 8 Sections 8.6 and 9.1
The theoretical premise or "school of thought" on which the work is based is challenged.	Ensure you have shown the diversity of theories (cite the literature) and demonstrate that you are testing one of these theories. If you are challenging accepted wisdom: • explain the theory being tested; • cite references which explore the theory; and • use writing structure (e.g. section headings, topic sentences) to stage the development of your logic.	Introduction (Stages 2 and 4; may be re-emphasized in the Discussion Introduction (Stage 2; may be emphasized in the Discussion)	Chapters 8 and section 9.1 Chapter 8
The experimental design or analysis methods are challenged.	Include constraints (conditions when the results may not apply). Defend the design or analysis on its merits. Refer to previously published examples using the design or analysis (cite the literature). Include additional information on the design or analysis if available.	Discussion In letter responding to referees' comments Methods; also in the response to referees' comments Methods (may be re-emphasized in the Discussion)	Chapter 9 Section 14.3 Chapter 7 and section 14.3 Chapters 7 and 9
You are asked to supply additional data or information that would improve the paper.	Supply the additional data if you can.	Results (may include other sections)	Chapters 5 and 6

(*Continued*)

Table 14.2 (*Continued*)

Comment type	Author response	Where in manuscript	Section to check in this book
	If you cannot supply the extra data and only minor changes are required, make your case to the editor for not including new material.	In letter responding to referees' comments	Section 14.3
	If major changes are required, consider rewriting the paper to make additional information unnecessary.	All sections	Chapter 4
You are asked to remove information or discussion.	Remove the information if you can do so without changing the story. You can ask a colleague to make suggestions on where to make cuts if the referee is not clear.	Where indicated (usually Introduction and/or Discussion)	Various
	If cuts would alter your story too dramatically, weigh up the positives in the editor's letter and referee reports and make your case to the editor for retaining the information.	In letter responding to referees' comments	Section 14.3
The conclusions are considered incorrect, weak, or too strong.	Ensure the Discussion is tied to the aims at the beginning of the paper.	Cross-check Discussion with Introduction (Stage 4)	Chapter 9 and section 8.6
	Reassess the literature you have cited and make a case to the editor if there is adequate supporting literature (check and cite supporting literature).	Discussion and in letter responding to referees' comments	Chapter 9 and section 14.3
	Ensure that all your statements are justified and the strength of language is appropriate.	Discussion	Section 9.2
	Include constraints (conditions when the results may not apply).	Discussion	Chapter 9
The referee has unspecific negative comments; e.g. poorly designed, written, or organized.	Show the referees' comments to a colleague and discuss them.	All relevant sections	Various
	Restate or rewrite the section(s) and note each change you make to the Editor.	In letter responding to referee's comments	Section 14.3
	Point out to the editor all the work you have done to improve the paper, i.e. build up a body of positives; e.g. "I have addressed point 1 by . . .".	In letter responding to referee's comments	Section 14.3
	Seek language or editorial assistance if the writing or grammar are criticized.	Relevant sections	Chapter 17

the comment is really about and consider whether the approaches recommended in Table 14.2 are still appropriate. If you have a conditional acceptance from the editor then none of the comments is enough to stop you publishing the paper. The main exercise now is to maintain the integrity of your story while accommodating the reviewers and editor.

Table 14.2 sets out each of the main types of comment you are likely to receive from the referees and editor and recommends a series of approaches to responding to the comments. The recommended responses to each comment type range from easy to more difficult, and some comment types may require a mixture of responses. Many reviewers' comments can be addressed by appropriate use of the two most powerful tools available to writers of scientific articles, as follows.

- **Citing the published literature**. Published works have already been reviewed and accepted by the scientific community. The findings and conclusions that have been published by different authors can be compared and contrasted and used to develop an idea or support an argument.
- **Improving the structure of the manuscript**. The structure and logic of each section and subsection of a scientific article are described in this book. Revising the relevant chapters of the book will help you to deal with reviewers' comments by helping you improve the structure of your ideas or arguments.

Use Table 14.2 to decide on the appropriate response(s) to comments and the place(s) in the manuscript where changes should be made (the reviewers' comments may also indicate where changes can be made). Table 14.2 also indicates which sections of this book to revise as part of dealing with reviewers' comments.

Task 14.1 Analyzing an authentic example

Ask a colleague who has had an article reviewed to show you the reviewers' comments and their response.

1 Decide which of the seven types of reviewer comments listed above were made.
2 Check whether the responses the author made fit the suggested response types in Table 14.2.
3 Discuss the thinking behind the responses with the author.

  See Chapter 16 for additional suggestions about using previous reviews as a training tool.

Return or re-submit your manuscript with a letter to the editor

It is important to respond quickly to reviewers' comments and the editor's recommendation about publishing the manuscript. This is true regardless of whether the manuscript has been accepted with minor changes or you have been encouraged to re-submit it after major revision. As with the covering letter you sent when you originally submitted the manuscript, the letter accompanying the revised manuscript is an opportunity to demonstrate that you appreciate the role of the editor and that you have done everything you can to improve the manuscript to meet the journal's and the reviewers' requirements. Use the letter responding to reviewers' comments to do these things:

85

How to respond to editors and referees

Ch 14
How to respond to
editors and referees

To: Dr AB Brown,
Editor, *Journal of*..................

Re: Manuscript Number.......
Title................................
Authors............................

Dear Dr Brown,

Thank you for your letter accepting the manuscript entitled ... pending revision. We have made all the changes you suggested in your letter and address all the comments of the two reviewers in the notes below. We have also attended to the formatting and language of the manuscript according to your suggestions. Please note that reviewer comments are shown in **bold** type and our responses in plain type.

We note that there was some disagreement between the reviewers about the usefulness of the section of manuscript on 'observer effects' and that only Reviewer #1 recommended that this section be dropped. We are concerned that omitting this section might contribute to a lack of transparency and repeatability. It is critical to deal with it, because without it our key result would be confounded. Also, in discussions with colleagues on this topic, observer effects are invariably a subject of keen interest, and we believe readers would be frustrated to have our approach to dealing with it relegated to a brief reference. We have made some minor changes to the 'observer effects' section to shorten it. We would be willing to make further changes if you felt them necessary and would be grateful for your advice on the matter.

(Continued)

Fig. 14.2 An adapted letter from an author to an editor, responding to reviewers' comments. (N.B. This is an unusually short letter.)

- list the main changes you have made individually, referring to referees' reports;
- say you have also corrected minor errors (e.g. English);
- point out supportive comments by referees and any disagreements between them (side with the reviewer you think is right and try to get the editor on your side);
- defend your work if a referee is factually wrong (another chance to cite key published papers supporting your argument); and
- say you believe the paper is important research and is now acceptable.

Copy all of the reviewers' comments into your letter (use a typeface that distinguishes them from your responses, e.g. bold) and write a response to each one. Re-check that the changes to the manuscript conform to the guidelines in the Instructions to Contributors (e.g. formatting, length, style). Figure 14.2 shows an example of a letter responding to reviewers' comments.

Send the revised manuscript back to the editor, together with your letter responding to the reviews.

Ch 14
How to respond to
editors and referees

Response to comments by Referee #1

1. Survey site markers in Fig 2 are too small.

Survey site markers have been increased in size.

2. How were $a_n(x)$ and $b_n(x)$ computed? If they were computed empirically this should be stated in the text.

Yes, $a_n(x)$ and $b_n(x)$ were computed empirically. The relevant section now reads: "The quantities $a_n(x)$ and $b_n(x)$ were derived empirically, by calculating, for each visit and both survey types, the proportion of patches in which x species had been seen by visit n. For example, after three different day surveys, there were eight patches in which 17 species had been discovered, so $a_3(17) = 8/38 = 0.21$."

3. The notation in the equations is very complex and as this paper may be of interest to practitioners it would be better to reduce the use of symbols in Equations (1)–(7).

The notation of Equations (1)–(7) comes from another paper, so must be left as is. However, we have eliminated the use of β in reference to statistical power, and just used the word 'power' instead.

Response to comments by Referee # 2

All suggested corrections made by Referee #2 have been made in the text.

We believe the paper is now acceptable for publication and look forward to your response to the changes we have made.

Yours sincerely,
Dr Zhu

Fig. 14.2 (*Continued*)

CHAPTER 15

A process for preparing a manuscript

There are many different ways to proceed towards preparing a manuscript for submission to a journal, but the process often seems to take a very long time and involve a considerable amount of back-tracking and reworking. Indeed, multiple drafts are a necessary part of manuscript writing – as co-authors make their respective contributions and the paper's story is refined and strengthened – but it is in everyone's interests to streamline the process as much as possible. Here we present a possible set of steps for you to consider.

15.1 Initial preparation steps

1 Select a "package" of results that you think will make a paper. Collect the relevant data and discuss with your potential co-authors issues such as these.

- What are the take-home messages from these data (what story do the data tell)?
- Is this the best package of data to concentrate on? Should more data be included to strengthen the story, or should some data be removed to ensure that a single, coherent story can be told?
- Who are the target audience for the paper, how significant is the story told by the data, and therefore which journal should be selected as the target?
- How will the work of preparing the manuscript be divided up (i.e. who will do what)?
- Who will be listed as authors, and in what order will their names be shown? Who should be acknowledged for assistance? (It might be helpful to consult a source such as the website developed by the International Committee of Medical Journal Editors for criteria to use in determining who qualifies as an author; www.icmje.org/index.html#author.)
- What timeline is feasible? At which stages will the co-authors read drafts? (Once a decision has been made about this issue, you can insert steps at relevant places in the list below that refer to responding to feedback from co-authors.)

2 Give a short talk to a small group of your colleagues and present some background and reasons for the research (Stages 2 and 3 of the Introduction); the

Writing Scientific Research Articles: Strategy and Steps, 1st edition. By M. Cargill and P. O'Connor. Published 2009 by Blackwell Publishing, ISBN 978-1-4051-8619-3 (pb) and 978-1-4051-9335-1 (hb)

aims or hypothesis; an outline of key methods; all the data needed to tell the story (all the figures, tables, and other text); and a discussion of the results and their meaning. Ask the group to provide feedback on anything which was not clear in your presentation and any questions they have about the research.

3 Obtain the Instructions to Contributors for the target journal and consider setting up a document template following the guidelines provided.

4 Refine the tables and figures that present the data (following the journal's guidelines). As part of this process, consider whether all the tables and figures are necessary to tell the story, and combine or delete as appropriate. Make sure the take-home message of each table and figure is clearly visible to a reader, and easy to identify from the text in the title or legend.

5 Draft the Results section, highlighting the take-home messages.

6 Write bullet points on possible ideas to include in the Discussion.

7 Draft or refine the paper's title to capture the paper's main message.

8 Draft the Methods section, or its equivalent.

9 Draft the Introduction. Consider writing the stages in the order 4, 3, 1, 2, with Stage 5 (if present) at the appropriate place for your particular story (see Chapter 8 for details).

10 Draft the Discussion section, and the Conclusion if it is to be present.

11 Draft an Abstract/Summary.

12 Draft a set of keywords.

13 Put all the pieces together to form a complete first draft.

14 To refine your manuscript, follow the suggestions in section 15.2.

15.2 Editing procedures

1 Put the completed draft aside for a while. The literature on this topic suggests that you need at least 48 hours away from the draft before you can read what you actually wrote, as opposed to what you think you wrote.

2 When you come back to the document, print off a paper copy and read the document through from the beginning with the aim of identifying places where content changes are needed. Don't stop to make any changes, just put marks in the margin or under problem words, to indicate the places you will need to return to later.

3 Once you have reached the end of the document, go back to the beginning. Work on improving each problem you identified.

4 Then edit it again, as before.

5 Do this as many times as necessary. When you have completed this part of the process, you should be satisfied with the science of what you have written.

6 Now edit for so-called discourse features: these are the language features that contribute to the flow and linking of the sections and sentences.

- Check that subheadings appear wherever they are needed.
- Check that paragraphs have topic sentences where appropriate.
- Check that paragraphs and sentences follow our guidelines on leading from the general to the particular and on giving old information before new (see Chapter 8 for details).

7 Edit for spelling, punctuation, and grammar.

- Check especially for the mistakes you often make: use the Find feature of your word processor.
- *Always* have the computer's spelling checker switched on, but remember its limitations and that it cannot identify where you have used a word that is correctly spelled but is not the correct one in the context; e.g. if you type *there* where you mean *their*, or *it's* where you mean *its*. You will also need to add to the program's dictionary all the technical terms you use (checking carefully that they are spelled correctly when you add them!). Then you can be confident that every time a red wiggly line appears there really is an issue to be addressed.
- Check for punctuation and italics, especially *et al.* and species names. (Different journals have different conventions about these issues, so make sure you check in the Instructions to Contributors to find out what applies in the journal where you will submit.)
- If you use English as an additional language, editing your own writing for grammatical accuracy needs special care. We suggest that you use a ruler and hard copy of the text (i.e. do not try to do this on the computer screen). Start with the *last* sentence of a section and lay the ruler under the sentence. Read the sentence and check its grammar; i.e. making sure that the verbs and subjects agree, that singular and plural forms are used appropriately, that the verb tense is correct, and that the articles (a/an/the) are used appropriately. Then move the ruler up the page and read the sentence before the one you just checked. In this way you are less likely to be distracted by issues other than the ones you are supposed to be looking for: the grammatical ones. Remember, you are already happy with the science of the manuscript, after completing Steps 1–5 above as many times as necessary! Now you are only focusing on the grammar.

8 Edit for the correctness and consistency of the referencing and the reference list.

- If you are using one of the commercially available bibliographic software programs, such as Endnote or Reference Manager, most of this step has been done for you, but you will still need to check that the output of the program appears as you want it and that no entries have been produced that have anomalies or inconsistencies, which can occur if data has been entered into the program incorrectly.
- If you have produced the reference list manually, you will need to check carefully for these three things.

  i Does every reference in the text have a corresponding entry in the list?
  ii Does every entry in the list appear at least once in the text?
  iii Do all references in the text and all entries in the list follow the style stipulated by the journal exactly (i.e. including punctuation, spacing, use of italic and bold fonts, and capitalization (the use of capital letters))?

9 Edit for layout: view each page singly using Print Preview to ensure that headings stay with the following text and running headers appear or not as stipulated in the Instructions to Contributors for your target journal.

10 Check that you have followed the formatting requirements as provided in the Instructions to Contributors, including in regard to the placement in the manuscript of tables and figures and their titles and legends, and the provision of any supplementary data to appear on an associated website, if applicable in your case.

11 Final check: do a final read-through to catch the "little" mistakes that may have slipped by. It can be very helpful to ask a colleague or friend to do this for you: remember also to make yourself available to do the same for them when their turn comes to submit a manuscript.

15.3 A pre-review checklist

Now you are ready to ask for some serious feedback on the article, from people outside the author team. One option for this step is to provide your critical reader with a list of questions to respond to. In Table 15.1 we provide such a list, which has been developed on the basis of the material covered in this book. An electronic version of the checklist is available on our website (www.writeresearch.com.au) so that you can easily download and adapt it to your specific purposes and the conventions of your own discipline.

Another option, perhaps to be used after the checklist, is to ask an experienced colleague to pre-review your manuscript; that is, to read it as if they were reviewing it for the journal. If appropriate you could provide them with the example Referee's Evaluation Form given in Figure 13.2.

Table 15.1 Checklist for review of paper drafts.

	Criterion	Reviewer's comments
1	Does the title reflect accurately the content of the paper?	
2	Are the significant words in the title near the beginning to catch a reader's attention?	
3	Does the Introduction begin with the big issue of topical/scientific interest and then narrow down to the specific topic of the paper?	
4	Does the Introduction locate the study effectively within the recent international literature in the field?	
5	Does the Introduction highlight a gap that the research fills, or present a need to extend knowledge in a particular area? (Does it say why the work was done?)	
6	Does the Introduction end with a clear statement of the aim/hypothesis of the research, or summarize the main activity of the paper (depending on the field and relevant journal conventions)?	
7	Are the methods, including statistical analysis, appropriate for the questions addressed and the study conducted?	
8	Are the materials and methods given in enough detail to convince a reader of the credibility of the results?	
9	Do the results provide answers to the questions raised in the Introduction, or fulfil the objectives given?	

10 Are the results presented in a logical order (either similar to the order of presenting the aims or methods, or similar to the order in which the Discussion is presented).

11 Are all the tables and figures needed to tell the story of the paper? Could any be combined or deleted?

12 Do all the tables and figures stand alone? (i.e. can readers understand them without going back to read the text of the paper?)

13 Does the Discussion begin with a reference to the original aim/hypothesis/question?

14 Are the results compared with other relevant findings from the literature? Are you aware of any other comparisons that could be made? Are appropriate explanations/speculations included about reasons for observed similarities, differences, and other outcomes?

15 Are appropriate statements made about the wider significance of the results, their limitations, and/or their implications for practice and/or future research directions?

16 Does the paper end with an appropriate concluding paragraph or section that emphasizes the key message(s) and their significance to the field?

17 Is the list of references complete (all the works in the list are referred to in the paper, and all the works referred to in the paper are in the list)?

18 Are the reference list and in-text references formatted accurately and in the right style for the target journal?

19 Does the Abstract include all the information required by the journal, and does it highlight appropriately the key results and their significance?

20 Does the Abstract adhere to the word limit and follow the prescribed format of the target journal?

21 Are the selected keywords those that will best allow the article to be located by the full range of its prospective readers?

22 What additional comments do you have for strengthening the paper?

Once you have responded to the feedback received in this way, and done a final check, you are ready to submit your manuscript. Good luck!

Developing your publication skills further

CHAPTER 16

Skill-development strategies for groups and individuals

A number of effective strategies and activities can be implemented within research groups, laboratories, or departments to provide a structure or focus for developing publication skills and capacity. At one end of the spectrum these can be organized by the senior scientists, with students and junior members encouraged or required to participate. At the other end, groups of students or early-career researchers can join together to set up activities they think will benefit their own development, and request input from the senior staff as appropriate.

If your group is located in a country where English is not the working language, then the extent to which these activities take place in English is a decision to be made on a case-by-case basis. It can be helpful to involve at the planning stage an English-teaching professional with relevant expertise, to discuss where and how English improvement can be built into the activities. Many of the sections of this book are candidate materials for structured input to these sessions, perhaps followed by a time for a discussion of someone's draft paper, or the slides for an upcoming conference presentation.

The following sections present some ideas for different types of activities that can be used. We recommend that any strategy be planned to have a limited duration (e.g. meeting every 2 weeks for 3 months, followed by a review), an agreed set of objectives, and explicit ground rules for how the sessions will run, preferably agreed by all the participants at the first meeting.

16.1 Journal clubs

A journal club is a popular strategy used in many science fields to build levels of knowledge in specific areas. It involves all members of the club reading the same journal article (nominated by the group leader or a designated group member) and then coming together to discuss it in depth. The discussion sessions are chaired by a member of the group (this role usually rotates among the membership), who is often expected to identify particular points within the article for focused discussion.

An additional component can be added to the end of these sessions to include a publication skill emphasis. Participants can be asked to analyse one of more of the

Writing Scientific Research Articles: Strategy and Steps, 1st edition. By M. Cargill and P. O'Connor. Published 2009 by Blackwell Publishing, ISBN 978-1-4051-8619-3 (pb) and 978-1-4051-9335-1 (hb)

article's sections (its title, Abstract, Discussion, etc.) using tasks from the relevant sections of this book. The aim would be to answer questions such as these.

- Is this section effective in terms of communicating this content with its intended audience?
- What makes it effective in your opinion?
- Can you find examples of the techniques highlighted in this book that contribute to the effectiveness?
- Can you identify additional features that make it effective?
- Can you identify anything that could be improved?

16.2 Writing groups

Writing group is a general name for any group of people who come together on a regular basis to enhance their progress on their individual writing projects: in this context, probably article manuscripts or thesis chapters. Writing groups can be facilitated (a more experienced person provides leadership or input) or unfacilitated (the group members run the group activities themselves). Both types can be useful, depending on the circumstances, work patterns, and learning-style preferences of the prospective participants.

At a basic level, two or three people can commit to meet on a regular basis to read each other's drafts, with an agreement made at the end of each meeting about who will provide a draft section to the others by an agreed date, for discussion at the next meeting. A helpful discussion of how to structure this type of group for best benefit, written by one who has done it, can be found at http://chronicle. com/jobs/news/2007/08/2007080801c/careers.html.

16.3 Selecting feedback strategies for different purposes

You may be asked to give feedback on another person's writing in the context of a writing group, as a personal request, or in a more formal capacity as a reviewer for a conference or a journal. As it is rare for training to be provided on the giving of feedback, we present below some comments for you to consider as you approach the task.

Before you give feedback on someone else's writing, it is helpful to clarify the role you have been asked to play. Writers often have a strong emotional investment in their writing, and they can sometimes feel under personal attack if they receive comments on it that do not fit with their views of the relationship between writer and reviewer and the role they expected the reviewer to take. So, when someone asks for your feedback, it can be helpful to discuss with the requester what type of feedback they are seeking and what role they want you to take in this particular instance.

One possible feedback type that can be requested is "just the content" or "just the science," with the requester not expecting comments on the language used to express the content. This request is very difficult to carry out for many reviewers; one way to do it is for the reviewer to use the checklist for review of paper drafts

(Table 15.1) without annotating the draft itself at all. Another way is for the writer to seek feedback on the main points of the content (the take-home messages or story of the paper) before the writing of the full manuscript is undertaken. This can involve reviewing the answers to the four key questions given in Task 4.1, plus the full set of tables and figures that provide the evidence for the story.

Once feedback is being sought on a full draft of a paper, it is most likely that comments will be forthcoming on all aspects of the text. In this situation, it is useful for the provider of the feedback to think about what role they will adopt, perhaps by reflecting on the questions below.

- To what extent is my purpose to coach (encourage and suggest ways of improving in a supportive way)?
- To what extent is my purpose to act as a gatekeeper (one who decides whether the work is good enough or appropriate for its purpose)?
- To what extent is my purpose to teach (focus on helping the writer learn things that will become part of that person's repertoire of skills for the future)?
- What other purposes do I have?

Once you have made some decisions about these points, it may be helpful to think about how much power you want to adopt in your relationship to the writer whose writing you are reading.

- Do you want to appear as an expert who definitely knows the answers and whose advice must be followed?
- Do you want to appear as a more experienced colleague who can suggest things on the basis of your experience, and whose advice should be seriously considered?
- Do you want to appear as a fellow struggler with the issues, someone who is also learning how to write for the international English-language literature, who can act as an example of the intended audience, and apply the learning from this book to make suggestions and see if the writer agrees with them?
- Do you want to blend these approaches, adopting more of one in some areas, such as the science, perhaps, and more of another for other aspects?

Will your answers to these questions change the words you use to provide written feedback on drafts? For example, in what circumstances would you be more likely to use each of the following options?

- More explanation needed.
- Not sure what you mean here.
- Move this to the Introduction.
- This may fit better in the Introduction.

In thinking more globally about your feedback style, it can be useful to consider which of the following feedback strategies you have used before, and which you would like to try in the future:

- commenting on aspects that have been well done before pointing out things to be improved;
- using different coloured ink for different categories of feedback (science and language, for example);
- restricting yourself to the most important issues: intentionally not correcting everything in the case of early drafts;
- at the end of the document, providing a summary of both the positive aspects and the most important changes you recommend;

- recommending other sources of help: other people to talk to, books, or electronic resources to consult;
- using a set of symbols such as those below to indicate the types of issues needing attention, instead of or as well as writing proposed corrections on the manuscript.

sp	= spelling
p	= punctuation
sing/pl	= wrong choice of singular or plural form
wo	= word order
agt	= agreement between subject and verb
t	= tense
art	= article (a/an, the, or no article)
obn	= put old information before new information

It is likely that your answers to all the questions above will depend on an even broader set of factors:

- your seniority (how much experience you have had);
- your institutional role (what your job requires you to do);
- your personality;
- your relationship with the requester (the author of the document you have been asked to comment on); and
- what the requester asked you to do.

Finding an appropriate balance in a given situation between all the possible ways of responding can be a challenge, but progress towards this goal can be extremely rewarding. In the end it contributes to a skill-set that is of considerable importance in the work of a publishing researcher: the ability to give feedback that is rigorous, constructive, and inclusive.

16.4 Training for responding to reviewers

To move beyond the necessarily general advice provided earlier in this book on this topic, we can suggest the following training strategy. It requires that one published member of your research group be willing to share with others the documents that represent the full process of getting one of their articles accepted for publication. In our experience this is usually a more senior member of the group who has an interest in developing the capacity of less experienced members. A suggested process for a training workshop (or a series of workshop meetings) is given below.

1 The author of the paper provides to each workshop participant copies of the originally submitted manuscript and the journal's initial response to it: the editor's letter and the referee reports.
2 Participants are asked to read these documents thoroughly. They then form small groups and discuss how they would have responded to the editor's and referees' comments.
3 Each small group shares their proposed responses with the large group. The author then describes what was actually done in response, and distributes

copies of the written response that was sent to the journal. It is helpful if the author includes here a description of the emotional response to the editor's letter that was felt by the corresponding author, and how those feelings were dealt with.

4   The small groups re-form. Participants read the response document, identify differences between it and their first ideas, and discuss possible reasons for them.

5   The large group reconvenes, and the author comments on the issues identified by each small group in Step 4.

6   If there was a second round of reviewing, the process can be repeated if there are new insights to be gained from doing so. Otherwise the author can just explain the final outcome.

7   Participants are asked to summarize what they have learned from the workshop, in terms of both strategies for preparing their own responses and points to pay attention to in the original writing and editing of the manuscript prior to submission.

8   In EAL contexts, it is useful if participants also take note of any useful sentences or expressions from the example responses discussed in Steps 4 and 6 that could be re-used in their own writing.

CHAPTER 17

Developing discipline-specific English skills

17.1 Introduction

It can be helpful to think of the English you need to write about your research as one English among many Englishes: the English of marine biology, for example, or the English of plant biotechnology. Therefore, to a certain extent, people new to a research field need to develop their discipline-specific English even if English is their first language. We have included the aspects of English usage that are of general interest for scientist authors in the previous chapters on writing each section of an article. This chapter, on the other hand, focuses on those aspects of English grammar and usage that are of particular relevance to science authors who use English as an additional language (EAL). We begin with a discussion of types of error and how they are likely to affect the perceptions of editors and reviewers. We then introduce two strategies that can be useful for developing discipline-specific English writing skills: the concept of sentence templates and a computer-based tool called ConcApp. We then focus on a selected range of features of scientific writing in English that we find present problems for many EAL science authors. We hope you will find something useful for addressing your own needs within these three different approaches.

17.2 What kinds of English errors matter most?

Communicating meaning clearly is the crucial factor in scientific writing. It is worth thinking for a moment about what aspects of writing in English might interfere most seriously with clear communication of meaning.

What do journal editors say?

> As long as the science is good and can be clearly understood, I don't worry too much about the English – I have copy editors who can fix that. (Personal communication, October 2005, editor of an Australian-based international journal)

Writing Scientific Research Articles: Strategy and Steps, 1st edition. By M. Cargill and P. O'Connor. Published 2009 by Blackwell Publishing, ISBN 978-1-4051-8619-3 (pb) and 978-1-4051-9335-1 (hb)

Task 17.1 Types of error

1 Discuss (or make individual notes): if you were an editor of an international journal published in English, what problems might you anticipate when editing articles submitted by scientists from EAL backgrounds?

2 Below we list some types of error that are often made by EAL writers. Complete Table 17.1 by writing the number of each error type in the appropriate column in terms of how often and/or how seriously you think that error affects the communication of meaning in science writing: rarely/slightly; sometimes/moderately; or often/seriously.

Table 17.1 Task 17.1, part 2: Exercise in assigning types of English language errors to three possible levels of effect on meaning.

Rarely/slightly affects meaning	Sometimes/moderately affects meaning	Often/seriously affects meaning

Error types to be used in completing Table 17.1.

1 Incorrect usage of singular/plural forms (e.g. all tea leaves sample were oven dried).

2 Over-complex/inaccurate grammatical structures (e.g. This may be due to lower pH hinders dissolution of soil organic matter and decreases total dissolved Cu concentration because of Cu-organic complex reducing.).

3 Non-agreement of verbs and subjects (e.g. the results of this study suggests that . . .).

4 Incorrect choice of preposition (e.g. similar with the results of other researchers).

5 Non-standard usage of the articles *a/an* and *the* (e.g. the accumulation of Cu in human body).

6 Non-standard selection of modal verbs (e.g. *would* versus *will*, *can* versus *could* or *may*).

7 Incorrect choice of part of speech (e.g. drought resistance varieties).

8 Non-conventional selection of tense (e.g. present tense to refer to results of the study being reported).

See the Answer pages for some comments on these error types and how they can be perceived by readers.

3 List these error types again under the following headings according to the priority you give to avoiding them in your own writing: high priority/ medium priority/low priority. See the Answer pages for ideas about what types of strategies can be effective for addressing each error type.

Although not all journals have the copy-editor option, it seems that the quality of the science is a primary concern across the board: see the following quotations, from the Elsevier online editors' forum (www.elsevier.com/wps/find/editors. editors/editors_update/issue10d, accessed 16 January 2008).

This is a long-standing problem. In the past it was solved to a large extent by detailed copy-editing of accepted papers. I became aware that this was apparently no longer being done when papers started appearing with ungrammatical titles.

For the researcher and for the reviewer, we should emphasize the scientific contents of their work. Language skills should not be the barrier.

The Authors may have important data, which is useful for the Community, and must be helped.

The key points appear to be these:

- good science is the most important thing; but
- the science needs to be clearly understandable.

Our suggestions for achieving this are to:

- write short sentences first (two clauses only) and join them later if needed; and
- aim to develop a repertoire of ways of expressing meanings that are useful in your discipline (a repertoire is a range of possibilities to choose from).

The following sections provide some ways to develop your repertoire.

17.3 Strategic (and acceptable!) language re-use: sentence templates

Recent research on EAL authors writing for publication in English has found that re-using language from other papers in the same field is a common strategy, but there is considerable discussion about when it is acceptable to re-use language, and when the practice crosses into what can be called "textual plagiarism" (Flowerdew & Li 2007). What seems clear is that for science writing there is a divide in the way people think about the content – the science – and the way they think about the language used to express the content. The originality of the work is seen mostly to reside in the content: the data and their analysis and interpretation. This situation differs somewhat from that pertaining to writing in the humanities and social sciences, where the language is seen to form the argument, and therefore the content of the writing. Nevertheless, the very clear convention in academic writing in English is that, to avoid the suspicion of plagiarism, authors should use their own words to paraphrase the findings or conclusions of other researchers, as well as citing the source of the information. The section below suggests a way in which EAL and other authors can be more confident about avoiding inappropriate language re-use, while still taking advantage of the effective writing of other authors to develop their own repertoires. This option involves the construction of sentence templates for later re-use. We do this by separating the structure or framework of a sentence from the so-called content chunks, the noun phrases.

To understand this concept first read the purpose statement below, from an article by Li et al. (2000) entitled "Water use patterns and agronomic performance for some cropping systems with and without fallow crops in a semi-arid environment of northwest China".

As part of a long-term research effort aimed at establishing a sustainable rainfed farming system in the semi-arid and sub-humid regions of northwest China, this paper presents a detailed study on the water use patterns and agronomic performance for some cropping systems with and without fallow crops in a semi-arid environment.

The objectives of this study were to: (1) determine the grain and aboveground biomass production and water-use efficiency of individual crops grown in the rotation; (2) analyze the seasonal and inter-annual patterns of soil water storage and utilization as well as water stress for the four major rotation crops such as winter wheat, corn, potato and millet; (3) determine the grain and aboveground biomass production and water-use efficiency for different rotation systems and evaluate the capacities of the rotation systems with and without fallow crops to utilize soil water storage in conjunction with seasonal precipitation; (4) establish whether the introduction of fallow crops into the wheat monoculture significantly influences the quantity of water stored in the soil that will be used by the subsequent wheat crop; and (5) discuss the characteristics of soil conservation for different rotation systems.

If we cross out all the noun phrases that relate just to this particular study, what remains is a series of frameworks that we call *sentence templates*.

As part of a long-term research effort aimed at ~~establishing a sustainable rainfed farming system in the semi-arid and sub-humid regions of northwest China~~, this paper presents a detailed study on ~~the water-use patterns and agronomic performance for some cropping systems with and without fallow crops in a semi-arid environment~~. The objectives of this study were to: (1) determine ~~the grain and aboveground biomass production and water-use efficiency of individual crops grown in the rotation~~; (2) analyze ~~the seasonal and inter-annual patterns of soil water storage and utilization as well as water stress for the four major rotation crops of winter wheat, corn, potato and millet~~; (3) determine ~~the grain and aboveground biomass production and water-use efficiency for different rotation systems~~ and evaluate ~~the capacities of the rotation systems with and without fallow crops to utilize soil water storage in conjunction with seasonal precipitation~~; (4) establish whether ~~the introduction of fallow crops into the wheat monoculture~~ significantly influences ~~the quantity of water stored in the soil that will be used by the subsequent wheat crop~~; and (5) discuss ~~the characteristics of soil conservation for different rotation systems~~.

The frameworks or templates would look like this (NP = noun phrase).

As part of a long-term research effort aimed at [NP1], this paper presents [NP2]. The objectives of this study were to: (1) determine [NP3]; (2) analyze [NP4]; (3) determine [NP5] and evaluate [NP6]; (4) establish whether [NP7] significantly influences [NP8]; and (5) discuss [NP9].

N.B. You would only use this template if it enabled you to express the meanings you were trying to make. To help you decide what sorts of meaning they might be, it is useful to list and characterize the noun phrases that you crossed out to make the template, as demonstrated in Table 17.2.

We suggest that you continue to identify relevant sentence templates for yourself, whenever you read a research paper for your work, in order to add to your repertoire. We suggest that you take an extra 10 minutes or so after you have read a paper for its content. Use this time to identify any useful sentence templates, and record them in a special file or notebook. It may be useful to organize these notes according to the section of the paper where the sentence template would be useful.

Table 17.2 Relevant characteristics of noun phrases (NP) for use in sentence templates.

Noun phrase	Characteristics
1 establishing a sustainable rainfed farming system in the semi-arid and sub-humid regions of northwest China	verb + ing + NP + in + [NP of location]
2 a detailed study on the water use patterns and agronomic performance for some cropping systems with and without fallow crops in a semi-arid environment	*a study* + on + NP + in + [NP of location]
3 the grain and aboveground biomass production and water-use efficiency of individual crops grown in the rotation	NP + of + [NP referring to features of study already introduced]
4 the seasonal and inter-annual patterns of soil water storage and utilization as well as water stress for the four major rotation crops of winter wheat, corn, potato and millet	NP + for + NP stating subjects of study
5 the grain and aboveground biomass production and water-use efficiency for different rotation systems	NP + for + NP stating subjects of study
6 the capacities of the rotation systems with and without fallow crops to utilize soil water storage in conjunction with seasonal precipitation	*the capacities of* [NP] to + verb + object
7 the introduction of fallow crops into the wheat monoculture	*the introduction of* + NP + into + NP
8 the quantity of water stored in the soil that will be used by the subsequent wheat crop	NP of measurement
9 the characteristics of soil conservation for different rotation systems	NP referring to type of conclusions expected from the study

---

Task 17.2 Drafting a sentence template for Stage 4 of an Introduction

1 Find the Introduction paragraph that contains the Stage 4 in each of the PEAs. To refresh your memory, Stage 4 is made up of the very specific sentences that present the purpose/objectives of the writer's study or outline its main activity or findings. What would the sentence templates look like? Draft them on a separate sheet of paper. Check your answer in the Answer pages.

2 Find the Stage 4 in your SA. If it is suitable as the basis of a sentence template, construct one from it. Look at the noun phrases in your SA purpose statement. List them and note down any characteristics that will help you if you want to re-use the template in the future.

17.4 More about noun phrases

Discipline-specific noun phrases make up a very important part of the writing you need to do about your research. Identifying and learning them accurately is a very useful strategy for improving your writing. Here we present some grammatical details about noun phrases, and highlight one area of common difficulty associated with them.

A *noun phrase* is a group of words that does not include a finite verb (i.e. does not include a verb with a subject), built up around a single headword. The headwords are italicized in the following examples:

- the *mechanisms* of salt marsh succession;
- *interactions* involving carbohydrates;
- the seasonal and inter-annual *patterns* of soil water storage and utilization.

Note that long noun phrases can be made up of several smaller noun phrases, often joined together with prepositions.

A special case: noun-noun phrases

This kind of noun phrase can cause problems for EAL writers, in our experience. An example of a *noun-noun phrase* is "resource availability." This phrase means "availability of resources." To shorten phrases like this, it is very common in scientific English for the second part (of resources) to be moved in front of the headword (availability). When this happens, the part that moves is always written in its *singular* form (resource) and the preposition is omitted. (It is rare to find a possessive form with an apostrophe in such cases in science writing.) Similarly, "carbohydrate interactions" means "interactions involving carbohydrates." Table 17.3 contains some more examples, taken from the PEAs.

A good way to remember this construction is the following example:

food *for dogs*   is   *dog* food

Using the noun phrase concept to read about unfamiliar areas of science

To summarize the section above, science writing is largely made up of sentence structures (templates), which are usable for many different areas of science, plus noun phrases, which are often specific to particular areas. Once you understand this concept, you will probably find it easier to read articles from areas of science with which you are not completely familiar. This is because you can skip over the

Table 17.3 Examples of noun-noun phrases from the PEAs.

Noun-noun phrase	Extended form of the phrase
propagule pressure	pressure exerted by propagules
invasion success	success of invasions
field work	work conducted in the field
urchin disturbances	disturbances caused by urchins
legume root nodules	nodules on the roots of legumes
bacteroid activity	activity by bacteroids
bacteroid iron acquisition	acquisition of iron by bacteroids
soybean homologue	homologue in soybeans

> ### Task 17.3 Unpacking noun-noun phrases
>
> Write down three noun-noun phrases commonly used in your research field. Next to each, unpack the phrase to explain what it actually means. For example
>
> crop traits = traits exhibited by crops
>
> Note the difference in the usage of singular and plural word forms in the two forms of the phrases. We suggest that you make a list of the noun-noun phrases you see used repeatedly in articles in your field, and learn them accurately, including whether the forms are singular or plural. This will help improve the accuracy of your writing considerably.

unfamiliar noun phrases on your first reading, just concentrating on the sentence structures and main meanings. Then you can identify which noun phrases recur frequently, and use a dictionary or website to find out their meanings, if you need to know them. This will depend on your reason for reading the article. If you need to understand more about the area of research and are new to it, then you will probably need to look up many noun phrases. If you are reading the article only to find one specific piece of information, perhaps about the use of a method, you will need to look up fewer noun phrases. As you make your decisions about which ones to look up, remember to identify the headword of each noun phrase first, as this is the most important part for the sentence meaning.

The noun phrase idea can also help you to complete exercises in this book that involve writing about areas of science that are unfamiliar to you. For example, for readers who are unfamiliar with molecular biology and plant physiology, the PEA by Kaiser et al. (2003) (Chapter 18) may be challenging to read. Skipping over the complex noun phrases and focusing on the sentence structures will enable you to more easily do the exercises and understand the main point we are trying to teach. Of course, the same is true for the other PEA, Britton-Simmons and Abbot (2008) (Chapter 19), for readers who are unfamiliar with marine biology studies.

17.5 Concordancing: a tool for developing your discipline-specific English

All languages contain words and phrases that are commonly associated with other words or phrases (e.g. theory and practice; genetically modified organisms; the effect of something on something else). These collocations (words that are commonly used together) can be identified and studied. If you want to identify and learn common collocations that are used in writing about your own research field, you need to study texts (examples of writing) specific to that field. In this section we introduce a type of software program that can help you do this in a systematic way: a concordancer.

What does a concordancer do?

A concordancer searches a group of texts (called a *corpus*) for all examples of a particular search item. It displays the results as lines of text across the screen, with the search term highlighted in the middle. Results can then be sorted according to what is on the left or right of the search term (and one, two, or three words away from the search term), and this can provide data for your language learning. If the corpus you search is specific to your research field, you can search it in this way to improve your use of discipline-specific English.

Below, we first provide an example of what can be learned from a concordancing search of a discipline-specific corpus (Task 17.4), and then explain how you can download a low-cost concordancing program called ConcApp from the Internet, and also construct your own discipline-specific corpus.

Using ConcApp software

ConcApp is a low-cost concordancing program developed by Chris Greaves and downloadable from the Internet at www.edict.com.hk/pub/concapp/. The program is small in size and easy to learn, yet can quickly perform the searches needed to answer EAL writers' questions about language usage.

Task 17.4 Getting familiar with concordancing

Look at the ConcApp search results below, obtained by searching for the term "soil" in a corpus of articles from the field of soil science. Then read the questions and answers that follow.

> to utilise existing available *soil* water, unlike the perennial gr
> es (4 g oven dry wt basis) of *soil* were weighed into 40 ml polypr
> required 9 kg P/ha, whereas a *soil* with a high P sorption capacit
> concentration by 1 mg/kg on a *soil* with a low P sorption capacity
> 00, it was expected that this *soil* would have consistently been t
> capacity (PBC), which is the *soil*'s capacity to moderate changes
> and buffering capacity of the *soil*-an attempt to test Schofield's
> nisms that are present in the *soil*-plant microcosm environment. T
> etermined in a growth-chamber *soil*-plant microcosm study. Nodding
> 84) Lime and phosphate in the *soil*-plant system. Advances in Agro
> a where crops rely heavily on *soil*-stored water accrued in summer
> fertility on these particular *soil*s. Although this aberration has
> over in a range of allophanic *soil*s amended with 14Clabelled gluc
> alues for 9 different pasture *soil*s, 6 and 12 months after P fert

Q1 Is soil countable, uncountable, or both in these examples?
A1 Both. Countable examples include "a soil with a high P sorption capacity" and "9 different pasture soils;" an uncountable usage can be seen in "samples of soil were weighed."

Some of these usages are different from those found in everyday English, where soil is always uncountable. From this example, you can get an idea of how a ConcApp search of a discipline-specific text collection can help you identify English usages that are specific to that discipline.

Q2 How many different ways is the word soil used in these examples?
A2 Quite a few! For example, as well as its usage as a countable and uncountable noun it is used in noun-noun phrases, both as the headword ("pasture soils") and as the adjective-equivalent ("soil water"); and in hyphenated adjectival constructions ("soil-stored water") and noun-noun phrases ("soil-plant microcosm").

Our suggestion is that you construct a corpus (meaning *body* in Latin, but in this case a special-purpose collection) of English language journal articles from your own discipline(s), so you can search it for the use of words or phrases you need in your scientific writing. This will provide data, on your own desktop, for your ongoing learning of the specific English phrases and expressions used in your discipline.

Making a corpus

To be most useful, a corpus needs to consist of documents from your own subdiscipline, and of the type you are aiming to write. For example, a useful corpus for EAL scientists wanting to write articles for international publication would be at least 10 published research articles in their particular field. Our suggestion is that the articles to be used for a corpus be selected or approved by supervisors or leaders of research groups, to ensure that

- they are from reputable journals in the field;
- they are well-written, by authors using English as a first language or at a comparable level;
- they cover a suitable range of subtopics within the field, to give a good range of language usage; and
- they cover the required range of types of writing (e.g. including or excluding review articles, as desired).

Preparing documents for a corpus

To be searchable by ConcApp, the texts must be saved in text-only format (ASCII). If the selected articles are available in Microsoft Word format (e.g. if the author is willing to provide the text in this format), saving the file as text-only files (.txt) is a straightforward operation. If you can download the articles in html format, then the same process is possible. In both cases, delete the tables and figures, the author biodata, and the reference lists before saving as .txt files. If the articles are in .pdf format, then a somewhat tedious set of steps needs to be followed: see below for details. All files should be placed in a single folder on your computer for ease of searching.

Copyright issues

Making a single copy for use with a concordancer is comparable to making a single copy for research use.

Training

Notes and a tutorial on how to use ConcApp are available from the website.

Preparing text in pdf format for concordancing

A copy/paste procedure must be followed to convert the text to a text-only format. Only the written part of the article is needed, so do not copy biodata, tables and figures, reference lists, or acknowledgements, and do not include the headers or footers on the pages. The conversion process may require some trial and error at the beginning.

- Download the file (if online).
- Open the file in Adobe Acrobat Reader.

- Select the Text tool (the T on the toolbar) or the Column select tool in that menu if the paper is in columns.
- Select as much text as you can without including unwanted items such as headers and footers, page numbers, tables/figures, or the reference list.
- Copy the text (Control + C).
- Open your word processor (such as Microsoft Word).
- Paste the text into a new document (Control + V).
- Repeat the steps of selecting, copying, and pasting until the whole paper is copied.
- Select the Save as... option from the File menu in your word processor.
- In the next window, choose Text only as the file format and name the file before saving.
- Edit the text file as necessary (see below).

Some pdf files have security measures embedded to stop copying. Nothing can be done with these files. If Copy or Paste functions will not work, this is the cause. Care must always be taken not to copy the headers, footers, and page numbers into the new file. We find that the easiest method in the long term is copying the text from one page or column, pasting into a word processor, then repairing the text so that it is restored to its original continuous flow (deleting unwanted spaces in sentences, etc.). This avoids copying the unwanted parts from the outset. The whole process seems tedious at first, but becomes an almost mechanical routine with practice.

Task 17.5 Practice with concordancing

Practice using the concordancer (or read carefully) to examine the texts in your corpus of journal articles in order to answer the following questions.

1 Do article authors begin sentences with "Also"?
2 What about "In addition"?
3 How else is "addition" used?
4 Do authors use "I" or "we"?
5 What constructions are used with the verb "affect"?
6 What verbs are used with the noun "role"? And what prepositions are used after this word?

Now, think of other searches that you could try. Additional ideas for using ConcApp are to be found on our website at www.writeresearch.com.au.

17.6 Using the English articles (a/an, the) appropriately in science writing

For many of you who use EAL, the problem of using articles appropriately has been a constant since your early days of English learning. You may have seen the rules explained in many different ways, and learned them over and over again. You may be wondering why we have chosen to discuss this issue again here. We have included a section on article usage precisely because it is so difficult to master, especially for EAL users whose home language does not contain articles, and because it is often highlighted by journal editors and referees as needing attention in submitted manuscripts.

Indeed, in our experience editors and referees who speak English as a first language, and who therefore learned article usage by immersion at their mother's knee, may have limited understanding of the complexity of this part of the English language system. This complexity is reflected in the fact that effective computer software to identify or correct article errors has not yet, to our knowledge, been developed. This lack reflects the degree to which the use of English articles with any noun phrase depends on the meaning of the phrase in its particular context in the sentence, especially whether the noun phrase is used there in a generic sense or a specific sense. This question (generic or specific) relates also to the problems of meaning that can occur when articles are used inappropriately. It is therefore with the generic/specific question that we begin our discussion of article use.

Generic noun phrases

Generic noun phrases refer to any – or all – members of a particular class or category of living things, objects, or concepts. There are four ways to write these generic noun phrases in English.

1 If the noun is *countable*, you can make it generic by writing it in its plural form and not using any article.

2 An alternative when the noun is *countable* is to make it generic by using its singular form with the article *a* or *an*.

> e.g. Healthy crops can contribute substantial cadmium to human diets.
> A healthy crop can contribute substantial cadmium to human diets.

3 When the noun you want to use is *uncountable*, you make it generic by omitting any article. (Remember: uncountable nouns never have a plural form.)

> e.g. Cadmium exists in soils in many forms.
> Manipulation of soil pH can be effective in managing Cd contamination.

4 English has another possible way of making generic noun phrases which you need to recognize. Sometimes, a singular countable noun carries the generic meaning when used with the definite article *the*. This is often used when referring to living things or common machinery or equipment. (It is usually also possible to substitute the plural form of the word without an article, also changing the verb to agree, of course.)

> e.g. The earthworm can be found in many types of soil. (*or* Earthworms can...)
> The computer has become an important tool for researchers. (*or* Computers have...)

N.B. For science writing in particular, it is important to remember that as long as you are talking about a noun as a concept or general class (any or all of them), the noun *stays generic* (i.e. you may have to unlearn the rule that says a noun is specific after it has been used once in a passage of writing).

Specific noun phrases

Specific noun phrases refer to particular, individual members of a class or category, rather than the class as a whole. The reader and the writer both know which one or ones of the noun are being referred to. This requires the use of *specific noun phrases*, which involve the definite article *the*. There are three different reasons why a specific noun phrase may be required, as described below.

Task 17.6 Generic noun phrases

In the first paragraph of the Introduction to the PEA by Kaiser et al. (2003), reproduced below, underline examples of generic noun phrases using both countable and uncountable nouns.

Legumes form symbiotic associations with N_2-fixing soil-borne bacteria of the *Rhizobium* family. The symbiosis begins when compatible bacteria invade legume root hairs, signalling the division of inner cortical root cells and the formation of a nodule. Invading bacteria migrate to the developing nodule by way of an 'infection thread', comprised of an invaginated cell wall. In the inner cortex, bacteria are released into the cell cytosol, enveloped in a modified plasma membrane (the peribacteroid membrane (PBM)), to form an organelle-like structure called the symbiosome, which consists of bacteroid, PBM and the intervening peribacteroid space (PBS; Whitehead and Day, 1997). The bacteria, subsequently, differentiate into the N_2-fixing bacteroid form. The symbiosis allows the access of legumes to atmospheric N_2, which is reduced to NH_4^+ by the bacteroid enzyme nitrogenase. In exchange for reduced N, the plant provides carbon to the nodules to support bacterial respiration, a low-oxygen environment in the nodule suitable for bacteroid nitrogenase activity, and all the essential nutritional elements necessary for bacteroid activity. Consequently, nutrient transport across the PBM is an important control mechanism in the promotion and regulation of the symbiosis.

Check your answers in the Answer pages.

1 The noun phrase is specific because the phrase is referring to shared or assumed knowledge of one particular referent (= the thing being referred to).

  e.g. In recent years the growth of desert areas has been accelerating in the world.

2 The noun phrase may be specific because the phrase is pointing back to old information already introduced to the reader.

  e.g. A pot experiment was conducted in an acid soil. The experiment showed . . .

3 The noun phrase is specific because the phrase is pointing forward to information that specifies which one or ones being referred to.

  e.g. The aim of this study was to investigate the effect of liming on Cd uptake.

N.B. It is worth noting that when the structure NP1 + of + NP2 is used, the first noun phrase will be specific (i.e. have *the* in front of it) about 85% of the time. It is therefore a good idea to always use *the* in this situation, unless you are very sure that the extended noun phrase (the two noun phrases joined with *of*) is generic for some reason.

Task 17.7 Specific noun phrases

Reread the Introduction paragraph from the PEA by Kaiser et al. (2003) and draw a square around each specific noun phrase. Discuss with a colleague why each one is specific.

Check your answers in the Answer pages.

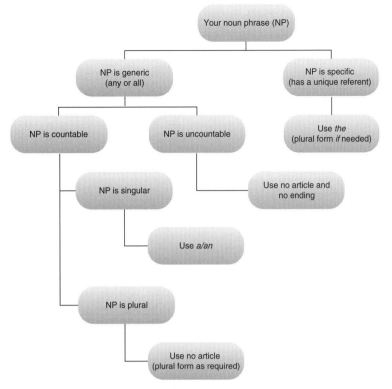

Fig. 17.1 Decision-support flowchart for the use of English articles (a/an/the) (after Weissberg & Buker 1990).

Summary flowchart for deciding on article use

Many EAL writers find the flowchart presented in Figure 17.1 helpful when they have to decide which form of the article to use with a noun phrase in a particular sentence.

Task 17.8 Articles and plurals in a science paragraph

Consulting the flowchart in Figure 17.1, fill in each blank space below with the plural marker -s, *a*, *an*, or *the* where necessary. (Some of the blanks do not require filling in.)

Propagule pressure
___ propagule pressure is widely recognized as ___ important factor that influences ___ invasion success. ___ previous studies suggest that ___ probability of ___ successful invasion increases with ___ number of propagules released, with ___ number of introduction attempts, with ___ introduction rate, and with ___ proximity to ___ existing populations of invaders. Moreover, ___ propagule pressure may influence ___ invasion dynamics after ___ establishment by affecting ___ capacity of ___ non-native species to adapt to their new environment. Despite its acknowledged importance, ___ propagule pressure has rarely been manipulated experimentally and ___ interaction of ___ propagule pressure with ___ other processes that regulate ___ invasion success is not well understood.

Check your answers in the Answer pages.

Problems with the relative pronouns *which* and *that*, and whether to use commas with them, are a common error we see in editing scientists' writing. The explanation below is designed to help you understand and remember how these two words are used.

> Example 1: Land *which is surrounded by water* is an island.

The italicized portion of Example 1 is the relative clause. In this example the relative clause is essential to the meaning of the sentence because if it were omitted the sentence would read "Land is an island". This does not make sense, because only land which is surrounded by water is called an island. Thus the relative clause defines which land the sentence refers to: it is a defining relative clause.

The important points to note about defining relative clauses are as follows.

- Practice differs in different parts of the English-speaking world in terms of the relative pronoun that can begin a defining clause. UK practice (which is also followed in Australia, New Zealand, etc.) allows either *which* or *that*. US practice (and the grammar checker in Microsoft software products) permits only *that* in this clause type.
- Defining clauses have no commas separating them from the rest of the sentence.

> Example 2: Tasmania, *which is surrounded by the waters of Bass Strait*, is an island of great natural beauty.

In this example, the relative clause is not essential to the basic meaning of the sentence. If it were omitted, the sentence would read "Tasmania is an island of great natural beauty" and this makes sense. The relative clause is adding extra, non-essential information and is thus a non-defining relative clause. Another way to work out if a relative clause is non-defining is to try inserting the phrase "by the way" after the *which*. If this addition sounds acceptable, the clause is non-defining.

The points to note about non-defining relative clauses are these.

- They are separated from the rest of their sentence by commas: two commas if they occur in the middle of the sentence as in our example, or one comma if they come at the end of a sentence.
- They can only begin with *which*.

N.B. The same distinction holds when the *which* + *verb* is omitted, forming a phrase.

> Examples: Tasmania, surrounded by the waters of Bass Strait, is an island of great
> natural beauty.
> Land surrounded by water is an island.

Task 17.9 Punctuation with which and that

Punctuate the following examples.

1 Lime which raises the pH of the soil to a level more suitable for crops is injected into the soil using a pneumatic injector.
2 Manipulation which involves adding or deleting genetic information is referred to as genetic engineering.

(Continued)

Task 17.9 (*Continued*)

3 Non-cereal phases which are essential for the improvement of soil fertility break disease cycles and replace important soil nutrients.

4 Senescence which is the aging of plant parts is caused by ethylene that the plant produces.

5 Opportunities that arise from the economically buoyant nature of domestic wine production must be identified and carefully assessed.

6 Seasonal cracking which is a notable feature of this soil type provides pathways at least 6 mm wide and 30 cm deep that assist in water movement into the subsoil.

7 Plants which experience waterlogging early in their development would be expected to have a much shallower root system than non-waterlogged plants.

8 Yellow lupin which may tolerate waterlogging better than the narrow-leafed variety has the potential to improve yields in this area.

9 Lucerne is a drought-hardy perennial legume which produces high-quality forage.

Check your answers in the Answer pages.

Provided example articles

CHAPTER 18

Provided example article 1:
Kaiser et al. (2003)

The Plant Journal (2003) **35**, 295–304

doi: 10.1046/j.1365-313X.2003.01802.x

The soybean NRAMP homologue, GmDMT1, is a symbiotic divalent metal transporter capable of ferrous iron transport

Brent N. Kaiser[1], Sophie Moreau[2], Joanne Castelli[3], Rowena Thomson[3], Annie Lambert[2], Stéphanie Bogliolo[4], Alain Puppo[2] and David A. Day[3,*]

[1]*School of Agricultural Sciences, Discipline of Wine & Horticulture, The University of Adelaide, Urrbrae, South Australia, Australia,*
[2]*Laboratoire de Biologie Végétale et Microbiologie, CNRS FRE 2294, Université de Nice-Sophia Antipolis, Parc Valrose, 06108 Nice cédex 2, France,*
[3]*Biochemistry & Molecular Biology, School of Biomedical & Chemical Sciences, University of Western Australia, Crawley, WA 6009, Australia, and*
[4]*Laboratoire de Physiologie des Membranes Cellulaires, UMR 6078 CNRS-Université de Nice-Sophia Antipolis, 284 chemin du Lazaret, 06230 Villefranche sur Mer, France*

Received 9 December 2002; revised 24 April 2003; accepted 7 May 2003.
*For correspondence (fax +61 08 9380 1148; e-mail dday@cyllene.uwa.edu.au).

Summary

Iron is an important nutrient in N_2-fixing legume root nodules. Iron supplied to the nodule is used by the plant for the synthesis of leghemoglobin, while in the bacteroid fraction, it is used as an essential cofactor for the bacterial N_2-fixing enzyme, nitrogenase, and iron-containing proteins of the electron transport chain. The supply of iron to the bacteroids requires initial transport across the plant-derived peribacteroid membrane, which physically separates bacteroids from the infected plant cell cytosol. In this study, we have identified *Glycine max divalent metal transporter 1* (*GmDmt1*), a soybean homologue of the NRAMP/Dmt1 family of divalent metal ion transporters. *GmDmt1* shows enhanced expression in soybean root nodules and is most highly expressed at the onset of nitrogen fixation in developing nodules. Antibodies raised against a partial fragment of GmDmt1 confirmed its presence on the peribacteroid membrane (PBM) of soybean root nodules. GmDmt1 was able to both rescue growth and enhance ^{55}Fe(II) uptake in the ferrous iron transport deficient yeast strain (*fet3fet4*). The results indicate that GmDmt1 is a nodule-enhanced transporter capable of ferrous iron transport across the PBM of soybean root nodules. Its role in nodule iron homeostasis to support bacterial nitrogen fixation is discussed.

Keywords: iron, NRAMP, nitrogen fixation, soybean, symbiosome.

Introduction

Legumes form symbiotic associations with N_2-fixing soil-borne bacteria of the *Rhizobium* family. The symbiosis begins when compatible bacteria invade legume root hairs, signalling the division of inner cortical root cells and the formation of a nodule. Invading bacteria migrate to the developing nodule by way of an 'infection thread', comprised of an invaginated cell wall. In the inner cortex, bacteria are released into the cell cytosol, enveloped in a modified plasma membrane (the peribacteroid membrane (PBM)), to form an organelle-like structure called the symbiosome, which consists of bacteroid, PBM and the intervening peribacteroid space (PBS; Whitehead and Day, 1997). The bacteria, subsequently, differentiate into the

N_2-fixing bacteroid form. The symbiosis allows the access of legumes to atmospheric N_2, which is reduced to NH_4^+ by the bacteroid enzyme nitrogenase. In exchange for reduced N, the plant provides carbon to the nodules to support bacterial respiration, a low-oxygen environment in the nodule suitable for bacteroid nitrogenase activity, and all the essential nutritional elements necessary for bacteroid activity. Consequently, nutrient transport across the PBM is an important control mechanism in the promotion and regulation of the symbiosis.

Micronutrients such as iron are essential for bacteroid activity and nodule development. The demand for iron increases during symbiosis (Tang *et al.*, 1990), where the

(a)

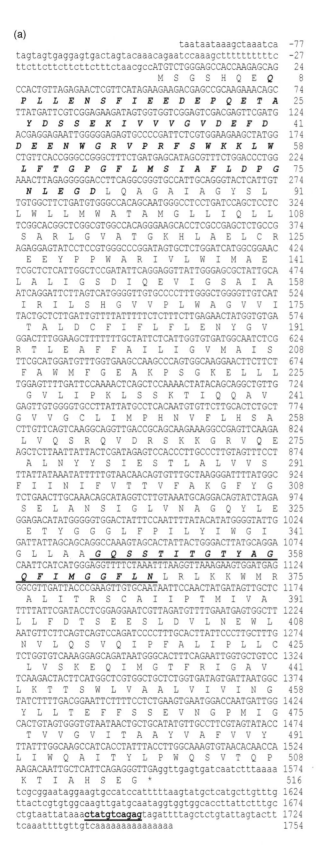

(b)

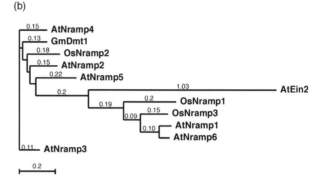

(c)

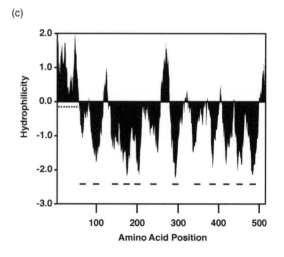

metal is utilised for the synthesis of various iron-containing proteins in both the plant and the bacteroids. In the plant fraction, iron is an important part of the heme moiety of leghemoglobin, which facilitates the diffusion of O_2 to the symbiosomes in the infected cell cytosol (Appleby, 1984). In bacteroids, there are many iron-containing proteins involved in N_2 fixation, including nitrogenase itself and cytochromes used in the bacteroid electron-transport chain. In the soil, iron is often poorly available to plants as it is usually in its oxidised form Fe(III), which is highly insoluble at neutral and basic pH. To compensate this, plants have developed two general strategies to gain access to iron from their localised environment. Strategy I involves secretion of phytosiderophores that aid in the solubilisation and uptake of Fe(III), while strategy II involves initial reduction of Fe(III) to Fe(II) by a plasma membrane Fe(III)-chelate reductase, followed by uptake of Fe(II) (Romheld, 1987). The mechanism(s) involved in bacteroid iron acquisition within the nodule have been investigated at the biochemical level, and three activities have been identified (Day et al., 2001). Fe(III) is transported across the PBM complexed with organic acids such as citrate, and accumulates in the PBS (Levier et al., 1996; Moreau et al., 1995), where it becomes bound to siderophore-like compounds (Wittenberg et al., 1996). Fe(III) chelate reductase activity has been measured on isolated PBM, and Fe(III) uptake into isolated symbiosomes is stimulated by Nicotinamide Adenine Dinucleotide (NADH), reduced form (Levier et al., 1996). However, Fe(II) is also readily transported across the PBM and has been found to be the favoured form of iron taken up by bacteroids (Moreau et al., 1998). The proteins involved in this transport have not yet been identified.

Two classes of putative Fe(II)-transport proteins (Irt/Zip and Dmt/Nramp) have been identified in plants (Belouchi et al., 1997; Curie et al., 2000; Eide et al., 1996; Thomine et al., 2000). The Irt/Zip family was first identified in *Arabidopsis* by functional complementation of the yeast Fe(II) transport mutant DEY1453 (*fet3fet4*; Eide et al., 1996). *AtIrt1* expression is enhanced in roots when grown on low iron (Eide et al., 1996), and appears to be the main avenue for iron acquisition in *Arabidopsis* (Vert et al., 2002). Recently, a soybean Irt/Zip isologue, GmZip1, was identified and localised to the PBM in nodules (Moreau et al., 2002). GmZip1 has been characterised as a symbiotic zinc trans-

porter, which does not transport Fe(II). The second class of iron-transport proteins consists of the Dmt/Nramp family of membrane transporters, which were first identified in mammals as a putative defence mechanism utilised by macrophages against mycobacterium infection (Supek et al., 1996; Vidal and Gros, 1994). Mutations in Nramp proteins in different organisms result in varied phenotypes including altered taste patterns in *Drosophila* (Rodrigues et al., 1995), microcytic anaemia (mk) in mice and belgrade rats (Fleming et al., 1997) and loss of ethylene sensitivity in plants (Alonso et al., 1999). The rat and yeast NRAMP homologues (DCT1 and SMF1, respectively) have been expressed in *Xenopus* oocytes and shown to be broad-specificity metal ion transporters capable of Fe(II), amongst other divalent cations, transport (Chen et al., 1999; Gunshin et al., 1997). The plant homologue, AtNramp1, complements the growth defect of the yeast Fe(II) transport mutant DEY1453, while other *Arabidopsis* members do not (Curie et al., 2000; Thomine et al., 2000). Interestingly, AtNramp1 overexpression in *Arabidopsis* also confers tolerance to toxic concentrations of external Fe(II) (Curie et al., 2000), suggesting, perhaps, that it is localised intracellularly.

In this study, we have identified a soybean homologue of the Nramp family of membrane proteins, GmDmt1;1. We show that GmDmt1;1 is a symbiotically enhanced plant protein, expressed in soybean nodules at the onset of nitrogen fixation, and is localised to the PBM. GmDmt1;1 is capable of Fe(II) transport when expressed in yeast. Together, the localisation and demonstrated activity of GmDmt1;1 in soybean nodules suggests that the protein is involved in Fe(II) transport and iron homeostasis in the nodule to support symbiotic N_2 fixation.

Results

Cloning of GmDmt1;1

A partial cDNA of GmDmt1;1 was identified from a 6-week-old soybean nodule cDNA library during a 5′-RACE PCR experiment designed to amplify the N-terminal sequence of a putative NH_4^+ transporter, GmAMT1. Subsequent PCR experiments identified a full-length 1849-bp cDNA, which was cloned and sequenced (Figure 1a) (accession no.

Figure 1. Sequence analysis.
(a) Nucleotide and the deduced amino acid sequence of *GmDmt1;1*. Amino acids italicised and in bold represent the N-terminal region of GmDmt1;1 used for the generation of the anti-GmDmt1;1 antisera. Consensus Dmt transport motif (bold italic underlined amino acids) and putative iron-responsive element (IRE; bold underlined) are indicated.
(b) Phylogenetic tree of selected members of the Dmt/Nramp family found in plants AtNramp1 (AF165125), AtNramp2 (AF141204), AtNramp3 (AF202539), AtNramp4 (AF202540), AtNRAMP5 (CAC27822), AtNramp6 (CAC28123), AtEin2 (AAD41076), OsNramp1 (S62667), OsNramp2 (AAB61961), OsNramp3 (AAC49720). The phylogenetic tree was drawn using MacVector (Accelrys) after comparison of deduced amino acid sequences using the CLUSTAL W method. The phylogram was built using the neighbour-joining method and best-tree mode. Distances between proteins were estimated using the Poisson-correction algorithm.
(c) Hydropathy analysis of the deduced amino acid sequence of GmDmt1;1 calculated using the Kyte and Doolittle algorithm with an amino acid window size of 19. Putative transmembrane spanning regions are indicated with horizontal bars. Dashed bar indicates hydrophilic section of protein used to generate anti-GmDmt1 antisera.

AY169405). Analysis of the GmDmt1;1 nucleotide sequence identified an open-reading frame of 516 amino acids encoding for a putative protein of approximately 57 kDa (Figure 1a). A BLAST search analysis of the GmDmt1;1 amino acid sequence identified significant homology (approximately 29% identity; approximately 46% similarity) to the amino acid sequences of six members of the *Arabidopsis* Nramp family (excluding AtEin2) of divalent metal ion transporters (Figure 1b). Hydropathy analysis (Kyte and Doolittle, 1982) of the encoded amino acids identified a protein with 12 putative transmembrane-spanning regions (Figure 1c). Between transmembrane segments 8 and 9, there is a conserved transport motif (5'-GQSSTITGTYAGQ-FIMGGFLN-3'), common among Nramp/Dmt homologues (Figure 1a). In the 3'-untranslated region of *GmDmt1;1*, there is an iron-responsive element (IRE) motif (5'-CTATGT-CAGAG-3') between bases 1688–1698 (Figure 1a).

A search of the Soybean TIGR Gene Index (http://www.tigr.org) yielded several soybean sequences similar to *GmDmt1;1*. These sequences consisted of expressed sequence tags (ESTs) aligned to make four tentative consensus sequences (TC84846, TC93163, TC94978 and TC82594), while a fifth sequence was from GenBank (accession no. AW277420). These partial sequences are between 65 and 98%, identical to *GmDmt1;1*. Sequence TC93163 has 98% identity with *GmDmt1;1* (isolated from cv. Stevens) and is likely to represent the same isoform from soybean cv. Williams. Obviously, *GmDmt1;1* is a member of a small gene family in soybean.

Gene expression

Northern blot analysis demonstrated that GmDmt1;1 is a nodule-enhanced protein. *GmDmt1;1* mRNA transcripts were abundant in nodules, but were only weakly detected in roots, leaves and stems (Figure 2a). Coincidently, nodule *GmDmt1;1* mRNA expression was the highest during the growth period, associated with maximum rates of symbiotic nitrogen fixation (20–40 days after planting), and decreased thereafter (Figure 2b,c). In young developing nodules, *GmDmt1;1* mRNA was barely detectable (Figure 2b).

Protein localisation

Antibodies were raised in rabbits against the N-terminal 73 amino acids of GmDmt1;1 (Figure 1c). This antiserum was used in Western blot analysis of 4-week-old total soluble nodule proteins, nodule microsomes, PBS proteins and PBM, isolated from purified symbiosomes. The anti-GmDMT1 antiserum identified a 67-kDa protein on the PBM-enriched nodule protein fraction (Figure 3a), but did not cross-react with soluble nodule proteins, PBS proteins or nodule microsomes (Figure 3a). Replicate Western blots incubated with pre-immune serum (Figure 3b) did not

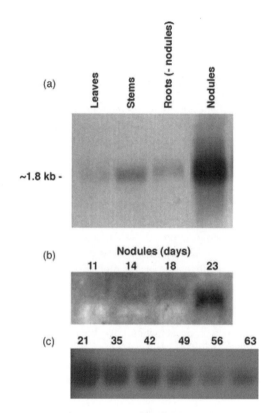

Figure 2. Northern blot analysis of *GmDmt1;1* expression.
(a) *GmDmt1;1* tissue expression. One microgram of poly(A)+-enriched RNA was extracted from 4-week-old soybean leaves, stems, roots (nodules detached) and nodules.
(b) GmDmt1;1 expression in developing nodules.
(c) GmDmt1;1 expression in mature nodules.
Ten micrograms of total RNA was extracted from the nodules prior to and after the onset of symbiotic nitrogen fixation. Blots (a) and (c) were probed with DIG-labelled antisense *GmDmt1;1* full-length RNA, while blot (b) was probed with randomly primed DIG-labelled full-length *GmDmt1;1* cDNA.

cross-react with the soybean nodule tissues examined. The protein identified on the PBM-enriched protein fraction is approximately 10 kDa larger than that predicted by the amino acid sequence of GmDmt1. The increase in size may be related to extensive post-translational modification (e.g. glycosylation) of GmDmt1, as it occurs in other systems. For example, the human Nramp1 and Nramp2 homologues are extensively modified by glycosylation and can appear about 40% larger on SDS–PAGE than predicted by their amino acid sequence alone (Gruenheid *et al.*, 1999; Tabuchi *et al.*, 2000, 2002). Post-translational modification of PBM proteins has been observed previously (Cheon *et al.*, 1994; Kaiser *et al.*, 1998), and the PBM protein Nod 24 undergoes extensive post-translation modification en route to the PBM, changing its apparent size on SDS–PAGE from 15 to 32 kDa (Cheon *et al.*, 1994). The localisation of GmDmt1;1 to the PBM was confirmed by subsequent immunogold-labelling experiments on fixed sections of infected cells containing symbiosomes. The anti-GmDmt1;1 antisera cross-reacted primarily with proteins on the PBM (Figure 3c,d).

Occasional cross-reactivity with bacteroids was also evident, but this was significantly reduced with more stringent blocking buffers, which included 5% w/v foetal albumin and 3% w/v normal goat serum (Figure 3e).

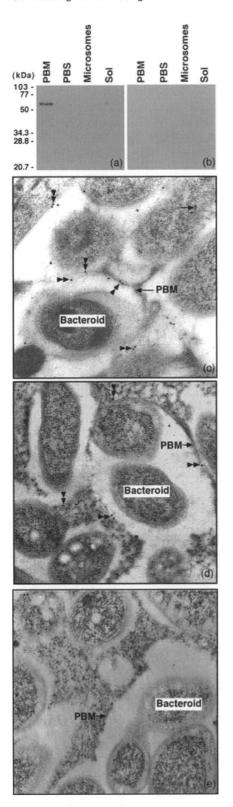

Functional analysis in yeast

To test for Fe^{2+}-transport activity, GmDmt1;1 and the positive control AtIrt1 (a known iron transporter) was cloned into the yeast-expression vectors, pFL61 and pDR195, and then transformed into the yeast iron-transport mutant DEY1453 (*fet3fet4*), which grows poorly on media containing low iron concentrations as a result of disrupted high (*fet3*)- and low (*fet4*)-affinity Fe^{2+}-transport activity (Dix *et al.*, 1994; Eide *et al.*, 1992). On synthetic-defined (SD) media supplemented with or without 2 μM $FeCl_3$, both AtIrt1 and GmDmt1;1 improved the growth of *fet3fet4* cells over those containing the empty cloning vector pFL61 (Figure 4a). Similarly, in liquid SD media supplemented with 20 μM $FeCl_3$ cells containing either AtIrt1 or GmDmt1;1 routinely entered the exponential-growth phase earlier than those of the empty vector controls (Figure 4b). In the absence of any added iron, GmDmt1;1 was unable to enhance growth of the mutant yeast (results not shown).

Short-term uptake experiments with 1 μM $^{55}FeCl_3$ showed that transformation of *fet3fet4* cells with GmDmt1;1 enhanced accumulation of ^{55}Fe(II) approximately fourfold over control cells (Figure 5a). This uptake followed Michaelis–Menten kinetics with an apparent K_M of 6.4 ± 1.1 μM (Figure 5b). The apparent K_M for Fe(II) agrees well with the need for supplementation of growth medium with micromolar iron in order to observe enhanced growth by the GmDmt1;1 cells (see above).

We tested whether GmDmt1;1 can transport other metal ions by heterologous expression in the zinc-deficient yeast-transport mutant, ZHY3 (*zrt1zrt2*) and the manganese transport mutant SMF1 (Chen *et al.*, 1999). On minimal zinc plates, GmDmt1 partially complemented ZHY3, but the growth of this mutant was slower than that of DEY1453 (*fet3fet4*) transformed with GmDmt1;1 (mean doubling times were 6.3 ± 0.5 h versus 5.1 ± 0.01 h ($n = 4$), respectively). In short-term transport studies, a 10-fold excess of $MnCl_2$ in the reaction medium inhibited ^{55}Fe uptake

Figure 3. Immunolocalisation of GmDmt1;1 to the peribacteroid membrane (PBM) of soybean nodules.
Western analysis of SDS–PAGE separated and blotted 4-week-old nodule protein fractions including enriched PBM, peribacteroid space (PBS) proteins, total nodule microsomes and soluble proteins. Duplicate blots were incubated with anti-GmDmt1;1 antiserum (a) or with pre-immune antisera (b) at a dilution of 1 : 3000, respectively. Thirty micrograms of purified protein was loaded in each lane. Molecular size markers are shown on the left. (c–e) Immunogold labelling of 3-week-old soybean nodule cross-sections of infected cells with symbiosomes. Tissue sections were incubated with anti-GmDmt1 antisera at a dilution of 1 : 100 (c, d) or with the pre-immune serum at a dilution of 1 : 50 (e) followed by 15-nm colloidal gold conjugated with goat antirabbit IgG (BIOCELL EM GAR 15) at a dilution of 1 : 40. Double arrows indicate immunoreactive proteins on the PBM and single arrows identify possible cross-contamination with bacteroids. EM magnification for both pictures was 35 000×.

significantly by DEY1453 (*fet3fet4*) transformed with GmDmt1;1 (Figure 5c). Similar inhibitions were seen with 10-fold CuCl$_2$ and ZnCl$_2$ (Figure 5c).

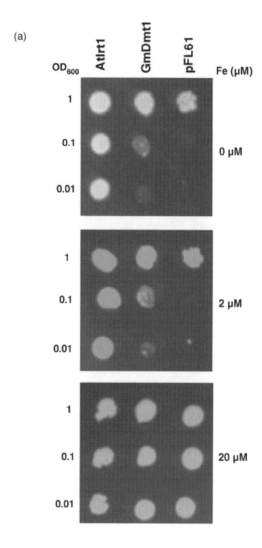

(a)

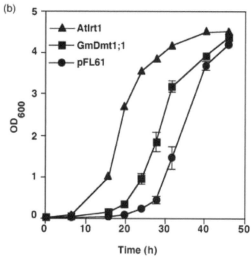

(b)

Discussion

GmDmt1;1 can transport ferrous iron

The results presented here demonstrate that GmDmt1;1 is a symbiotically enhanced homologue of the Nramp family of divalent metal ion transporters. The sequence of *GmDmt1;1* shares several common features with other members of the family, including 11–12 predicted transmembrane domains, a consensus transport motif between transmembrane domains 8 and 9 and an IRE in the 3'-UTR of the transcript (Gunshin *et al.*, 1997). Its expression is strongly enhanced in nodules, and immunological studies clearly localise the protein to the symbiosome membrane of infected cells. Its ability to rescue growth of the *fet3fet4* yeast mutant on low iron medium makes GmDmt1;1 a strong candidate for the ferrous iron transporter, previously identified in isolated symbiosomes from soybean (Moreau *et al.*, 1998). The kinetics of $^{55}Fe^{2+}$ uptake into complemented yeast (with an apparent K_M of 6.4 μM) also resemble those observed in isolated symbiosomes (linear uptake was observed over the range of 5–50 μM iron; Moreau *et al.*, 1998).

Specificity of GmDmt1;1

The competition experiments shown in Figure 5(c) indicate that GmDmt1 can transport other divalent cations in addition to ferrous iron. Zinc, copper and manganese all inhibited iron uptake. The ability of GmDmt1;1 to enhance growth of the *zrt1zrt2* yeast mutant further suggests that the protein is not specific for iron transport. The preferred substrate *in vivo* may well depend on the relative concentrations of divalent metals in the infected cell cytosol. This lack of specificity has been found with Nramp homologues from other organisms, including Nramp2 from mice. Despite this lack of specificity when expressed in heterologous systems, mutation of murine Nramp2 results in an anaemic phenotype, demonstrating that *in vivo* it is predominantly an iron transporter (Fleming *et al.*, 1997). Although GmDmt1;1 was able to complement the DEY1453 (*fet3fet4*) yeast mutant, the complementation was not robust and the growth media had to be supplemented with low concentrations of iron. AtIrt1, on the other hand, showed much better complementation and allowed growth of the mutant in the absence of added iron

Figure 4. Functional analysis of GmDmt1;1 activity in yeast cells.
fet3fet4 yeast cells were transformed with GmDmt1;1 inserted in the expression vector pFL61. Cells were also transformed with empty yeast expression vectors.
(a) Growth of serially diluted cells after 6 days at 30°C of GmDmt1;1 (GmDmt1;1-pFL61), AtIrt1 (AtIrt1-pFL61) and control (pFL61) transformed *fet3fet4* cells on synthetic-defined (SD) media supplemented with 0, 2, 20 μM FeCl$_3$.
(b) Growth in liquid SD media supplemented with 20 μM FeCl$_3$.

(Figure 4). There are several possible reasons for the poorer growth with GmDmt1;1, including possible instability of GmDmt1;1 transcripts (perhaps because of the presence of the regulatory IRE element in the transcript).

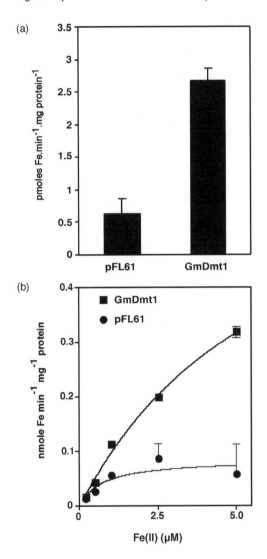

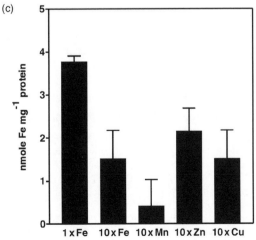

Localisation and function of GmDmt1;1

It has been suggested that AtNramp has an intracellular localisation (Grotz and Guerinot, 2002). The symbiosome is a vacuole-like structure (Mellor, 1989) and contains high concentrations of non-heme iron (Wittenberg *et al.*, 1996). However, this raises an interesting question as to the mechanism of GmDmt1;1. Divalent metal transport into vacuoles is likely to occur as Fe^{2+}/H^+ exchange (Gonzalez *et al.*, 1999), and it is possible that this also occurs in symbiosomes, as the PBM is energised by a H^+-pumping ATPase, which generates a membrane potential positive on the inside (and an acidic interior if permeant anions are present; Udvardi and Day, 1997). However, in this situation, and also in yeast, GmDmt1;1 catalyses uptake of iron into the cell, while uptake into symbiosomes is equivalent to export from the plant cytosol. Assuming that GmDmt1;1 is located in the plasma membrane of yeast and that it has the same physical orientation as in symbiosomes, which is likely considering that the secretory pathway is thought to mediate protein insertion into the PBM, then GmDmt1;1 must be able to catalyse bidirectional transport of iron. This is not unusual for a carrier and has been observed with GmZip1, a zinc transporter on the PBM. It appears that iron uptake can be linked to the membrane potential or pH gradient via other ion movements in the heterologous system. Further experiments on symbiosomes and yeast (or *Xenopus* oocytes) may provide new insights into the mechanism of iron transport in plants, but it appears that GmDmt1;1 has the capacity to function *in vivo* as either an uptake or an efflux mechanism in symbiosomes. This also raises the question of the relationship between GmDmt1;1 and the NADH-ferric chelate reductase on the PBM (Levier and Guerinot, 1996).

At the plant plasma membrane, ferrous iron transporters (presumably Atlrt1 homologues) act to take up iron reduced by the reductase into the plant. In the symbiosome, assuming that the orientation of the reductase on the PBM is similar to that on the plasma membrane, ferric iron stored in the symbiosome space would be reduced upon oxidation

Figure 5. Uptake of Fe(II) by GmDmt1 in yeast.
(a) Influx of $^{55}Fe^{2+}$ into yeast cells transformed with GmDmt1;1. *fet3fet4* cells were transformed with GmDmt1;1-pFL61 or pFL61 and then incubated with 1 μM $^{55}FeCl_3$ (pH 5.5) for 5- and 10-min periods. Data presented are means ± SE of ^{55}Fe uptake between 5 and 10 min from three separate experiments (each performed in triplicate).
(b) Concentration dependence of ^{55}Fe influx into *fet3fet4* cells transformed with GmDmt1;1-pFL61 or pFL61. Data presented are means ± SE of ^{55}Fe uptake over 5 min ($n = 3$). The curve was obtained by direct fit to the Michaelis–Menten equation. Estimated K_M and V_{MAX} for GmDmt1;1 were 6.4 ± 1.1 μM Fe(II) and 0.72 ± 0.08 nM Fe(II) min^{-1} mg^{-1} protein, respectively.
(c) Effect of other divalent cations on uptake of $^{55}Fe^{2+}$ into *fet3fet4* cells transformed with pFL61-*Gm*DMT1;1. Data presented are means ± SE of ^{55}Fe (10 μM) uptake over 10 min in the presence and absence of 100 μM unlabelled Fe^{2+}, Cu^{2+}, Zn^{2+} and Mn^{2+}.

of NADH in the plant cytosol. In isolated symbiosomes, addition of NADH together with ferric citrate, stimulated iron accumulation in the bacteroid, suggesting that the ferrous iron produced in the symbiosome space was taken up by the bacteroid ferrous iron transporter (Moreau *et al.*, 1998). *In vivo*, however, Fe(II) in the symbiosome space could also be transported back into the plant cytosol by the action of GmDmt1;1. We attempted to demonstrate this with isolated symbiosomes by loading them with $^{55}Fe^{3+}$ citrate, adding NADH and ATP (the latter to energise the membrane), and measuring efflux of ^{55}Fe into the reaction medium, but could not detect any efflux (Thomson, data not shown). The direction of transport *in vivo* will depend on the concentration of other ions on either side of the PBM and the activity of the bacteroid ferric and ferrous transporters.

Regulation of GmDmt1;1 expression

As mentioned above, GmDmt1;1 contains an IRE in its 3′-UTR. IREs are conserved sequences in the UTR of certain RNA transcripts to which iron-regulating proteins (IRPs) bind. The presence of an IRE motif suggests that *GmDmt1;1* mRNA may be stabilised by the binding of IRPs in soybean nodules when free iron levels are low. In both mammals (Canonne-Hergaux *et al.*, 1999) and *Arabidopsis* (Curie *et al.*, 2000; Thomine *et al.*, 2000), the abundance of Dmt isoforms containing an IRE element is enhanced by iron deficiency. Iron is required for both plant and bacterial enzymes during nodule development and in the functioning of the mature nodule. *GmDmt1;1* transcripts were detectable in relatively young (11-day-old) nodules and increased as the nodules matured (Figure 2). It is possible that during this time, when the bacteroid and plant iron requirements are relatively high, free iron levels are low and *GmDMT1* transcripts are stabilised by IRPs. This process could ensure nodule iron transport capacity through increased expression and activity of GmDMT1.

Conclusion

We have identified an Nramp homologue, GmDmt1, which is expressed in soybean nodules and encodes a divalent metal ion transporter located on the symbiosome membrane. The ability of this protein to transport ferrous iron makes it a candidate for the ferrous transport activity previously demonstrated in isolated symbiosomes (Moreau *et al.*, 1998).

Experimental procedures

Plant growth

Soybean (*Glycine max* L. cv. Stevens) seeds were inoculated at planting with *Bradyrhizobium japonicum* USDA 110 and grown in river sand in either glass houses under ambient light between 20 and 30°C, or in controlled-temperature growth rooms at 25°C day and 21°C night temperatures. Plants in the growth chambers were provided with a scheduled (14-h day/10-h night) artificial light (approximately 300 photosynthetic active radiation (PAR) at pot level) period. Plants were irrigated daily with a nutrient solution lacking nitrogen (Delves *et al.*, 1986).

Isolation of GmDmt1;1

Poly(A)+ mRNA was extracted from 6-week-old nodules (Kaiser *et al.*, 1998) and was used to synthesise an adaptor-ligated RACE cDNA library (Clontech; Marathon, Roche, Australia). A 480-bp cDNA amplicon was identified fortuitously from a 5′-RACE PCR experiment using an adaptor-specific primer, AP1: 5′-CCATCC-TAATACGACTCACTATAGGGC-3′ and GmAMTR24: 5′-CGAAC-CAAAGCATGAAGGTCCC-3′, a gene-specific primer designed against a partial cDNA of a soybean high-affinity NH_4^+ transporter, GmAMT1 (Kaiser, unpublished results). To amplify the complete *GmDmt1;1* cDNA, PCR experiments were performed using a second 6-week-old nodule cDNA library, which was ligated into the yeast-expression vector pYES3 (Kaiser *et al.*, 1998). Using primers pYES11R: 5′-GCCGCAAATTAAAGCCTTCG-3′ and GmDMTF2: 5′-AAGAATAAGGTGCCACCACC-3′, a 1.4-kb cDNA was amplified, which included the 3′-terminus of GmDMT1. A full-length clone (1.88 kb) was then subsequently amplified by the PCR from an adaptor-ligated 4-week-old nodule cDNA library (Clontech; Marathon) using high-fidelity Taq DNA polymerase (Roche) and primers AP1 and GmDMT1R21: 5′-AAAATTTGAAAGTACTAATACAGAGC-3′. Both strands of the full-length cDNA were sequenced.

Northern analysis

Total RNA was extracted from frozen soybean nodules roots after nodules were detached, stems and leaves using either a Phenol/Guanidine extraction method (Kaiser *et al.*, 1998) or the Qiagen RNAeasy system (Qiagen, Australia). Poly(A)+ RNA was isolated from total RNA pools using Oligotex resin (Qiagen). Ten micrograms of total RNA or 1 μg of Poly(A)+-enriched RNA was size-separated on a denaturing 1X MOPS 1.2% (w/v) agarose gel containing formaldehyde (Sambrook *et al.*, 1989) and blotted overnight onto Hybond N+ nylon membrane in 20× SSC. RNA was fixed to the membrane by baking at 120°C for 30 min. Blots were hybridised with either a full-length DIG-labelled antisense *GmDmt1;1* RNA produced using the SP6/T7 RNA DIG-labelling kit (Roche) or full-length randomly primed DIG-labelled *GmDmt1;1* cDNA. Blots were hybridised overnight at 68°C in DIG-easy hybridisation buffer (Roche). After hybridisation, the blots were washed twice for 15 min in 2× SSC, 1% SDS at ambient temperature, twice at 68°C for 30 min in 0.1× SSC, 1% SDS and twice for 15 min at ambient temperature in 0.1× SSC, 0.1% SDS, followed by chemiluminescent detection of the digoxygenin label using CDP-STAR (Roche).

Antibody generation and Western immunoblot analysis

To generate an antibody to GmDmt1;1, a 236-bp DNA fragment coding for 79 N-terminal amino acids was amplified using the PCR, using primers 5′-TGGCTCGAGCCACCAAGAGCAGCCACT-3′ and 5′-ACCCGAATTCCTGAAGGTCCCCCTCTAAG-3′. The DNA fragment was cloned into pGEMT (Promega, Madison, WI, USA) and was sequenced. The N-terminal DNA fragment was then subcloned into pTrcHisB (Invitrogen, San Diego, CA, USA) in-frame with the Histidine$_{(6)}$-tag and the initiation and termination codon. The resulting construct, pHISDMT1, was transformed into *Escherichia coli* TOP10F' cells (Invitrogen) and grown in 500 ml of liquid Solution B (SOB) media containing 50 µg ml^{-1} ampicillin at 37°C to an OD_{600} of 0.5. Expression of the $His_{(6)}$-tag GmDmt1;1 fusion protein was then induced by adding 1 mM isopropyl β-D-thiogalactopyranoside (IPTG) and incubating further for 3 h. Cells were collected and lysed in buffer (8 M urea, 50 mM NaH_2PO_4, 300 mM NaCl, 1.5 mM imidazole pH 8.0) and disrupted by six cycles of freezing and thawing followed by repeated passage through an 18-gauge needle. Insoluble proteins and cell debris were removed by centrifugation for 10 min at 16 000 *g*, and the supernatant was collected. The $His_{(6)}$-tagged GmDmt1;1 fusion protein was purified by immobilised metal affinity chromatography (Clontech, San Diego, CA, USA). Eluted protein was concentrated by tricholoracetic acid precipitation and re-suspended in elution buffer containing 8 M urea. The concentrated fusion protein (approximately 200 µg) was mixed with an equal volume of complete Freunds adjuvant (Sigma, USA) and injected into New Zealand White rabbits followed by four subsequent 200-µg injections at 1-month intervals. Ten days after the final injection, crude serum was collected. Protein fractions for Western immunoblot analysis were separated by 12 or 15% w/v SDS–PAGE (Laemmli, 1970) and blotted onto Polyvinylidene Fluoride (PVDF) membranes (Amersham, Buckinghamshire, UK), using a wet-blotting system (Bio-Rad, Regents Park, Australia). Membranes were probed with antiserum to GmDmt1;1 at a dilution of 1 : 3000 in PBS buffer, followed by secondary probing with a horseradish peroxidase-conjugated antirabbit IgG antibody. Immunoreactive proteins were visualised by chemiluminescence using a commercial kit (Roche, Australia).

Symbiosome isolation and nodule membrane purification

Symbiosomes were purified from soybean nodule extracts as described before (Day *et al.*, 1989), using a 3-step Percoll gradient. PBM-enriched membrane fractions were purified by rapid vortexing (4 min) of symbiosomes in buffer (350 mM mannitol, 25 mM MES-KOH (pH 7.0), 3 mM $MgSO_4$, 1 mM PMSF; 1 mM pAB; 10 µM E64; 1 mM DTT), followed by centrifugation at 10 000 *g* for 10 min in a SS34 rotor (4°C). The supernatant was collected and centrifuged further at 125 000 *g* for 60 min to separate the PBS proteins from the insoluble PBM-enriched membrane fraction. The PBM pellet was phenol-extracted (Hurkman and Tanaka, 1986), and the PBM and PBS fractions were concentrated by ammonium acetate/methanol precipitation and re-suspended at room temperature in loading buffer (125 mM Tris pH 6.8, 4% w/v SDS, 20% v/v glycerol, 50 mM DTT, 20% v/v mercaptoethanol, 0.001% w/v bromophenol blue). Soluble and insoluble nodule fractions were prepared by grinding nodules in buffer (25 mM MES-KOH pH 7.0, 350 mM mannitol, 3 mM $MgSO_4$, 1 mM PMSF, 1 mM pAB; 10 µM E64), followed by filtration through four layers of miracloth (Calbiochem, San Diego, CA, USA), and were centrifuged at 10 000 *g*, 4°C for 15 min to separate the bacteroids from the plant fraction. The supernatant was centrifuged further at 125 000 *g*, 4°C for 1 h. The supernatant was collected and concentrated by ammonium acetate/methanol precipitation. The nodule total membrane pellet and soluble protein fractions were re-suspended in loading buffer as described above.

Functional expression in yeast

GmDmt1;1 was cloned into the *Not*I site of the yeast–*E. coli* shuttle vector pDR195 downstream of the P-type ATPase promoter PMA1 (Thomine *et al.*, 2000) or into pFL61 under the control of the phosphoglycerate kinase promoter (Minet *et al.*, 1992). Yeast strain DEY1453 (*fet3fet4*) (Eide *et al.*, 1996) (*MATa/MATα ade2/+can1/can1 his3/his3 leu2/leu2 trp1/trp1 ura3/ura3 fet3-2::HIS3/fet3-2::HIS3/fet4-1::LEU2/fet4-1::LEU2*) was transformed (Gietz *et al.*, 1992) and selected for growth on SD media containing 20 mg ml^{-1} glucose and appropriate autotrophic requirements (pH 4.5; Dubois and Grenson, 1979). The media was also supplemented with 10 µM $FeCl_3$ to aid in the growth of *fet3fet4*. Yeast-uptake experiments were performed based on the protocol of Eide *et al.* (1992). fet3fet4 cells transformed with expression plasmids were grown to log phase in SD media with 2 µM additional $FeCl_3$. Log-phase cells were harvested, washed in H_2O and diluted in new SD media to an OD_{600} of 0.3 and grown for a further 4 h. Cells were harvested and washed twice with cold MES Glucose Nitriso-acetic acid (MGN) uptake buffer (10 mM MES, pH 5.5, 2% (w/v) glucose, 1 mM nitrilotriacetic acid). Cells were equilibrated at 30°C for 10 min before addition of an equal volume of $^{55}Fe^{2+}$ solution (MGN buffer, with 10 µM $FeCl_3$, $^{55}FeCl_3$ and 200 µM ascorbic acid to ensure that iron is in the ferrous form). Cells were incubated at 30°C, and aliquots were taken, filtered and washed five times with 500-µl ice-cold synthetic seawater medium (SSW) (1 mM EDTA, 20 mM trisodium citrate, 1 mM KH_2PO_4, 1 mM $CaCl_2$, 5 mM $MgSO_4$, 1 mM NaCl (pH 4.2)). Duplicate experiments were performed on ice as a background control for iron binding to cellular material. Internalised $^{55}Fe^{2+}$ was determined by liquid scintillation counting of the filters. Protein amounts were determined using a modified Lowry assay (Peterson, 1977).

Acknowledgements

This research was financially supported by a grant from the Australian Research Council (D.A. Day), the CNRS Programme International de Cooperation Scientifique, Program 637 (S. Moreau, A. Puppo) and a Canadian National Science and Engineering Research Council Postdoctoral fellowship (B.N. Kaiser). We thank Ghislaine Van de Sype for expert technical assistance with the microscopy.

References

Alonso, J.M., Hirayama, T., Roman, G., Nourizadeh, S. and Ecker, J.R. (1999) EIN2, a bifunctional transducer of ethylene and stress responses in *Arabidopsis. Science*, **284**, 2148–2152.

Appleby, C.A. (1984) Leghemoglobin and rhizobium respiration. *Annu. Rev. Plant Physiol.* **35**, 443–478.

Belouchi, A., Kwan, T. and Gros, P. (1997) Cloning and characterization of the OsNramp family from *Oryza sativa*, a new family of membrane proteins possibly implicated in the transport of metal ions. *Plant Mol. Biol.* **33**, 1085–1092.

Canonne-Hergaux, F., Gruenheid, S., Govoni, G. and Gros, P. (1999) The Nramp1 protein and its role in resistance to infection and macrophage function. *Proc. Assoc. Am. Physicians*, **111**, 283–289.

Chen, X.Z., Peng, J.B., Cohen, A., Nelson, H., Nelson, N. and Hediger, M.A. (1999) Yeast SMF1 mediates H$^+$-coupled iron uptake with concomitant uncoupled cation currents. *J. Biol. Chem.* **274**, 35089–35094.

Cheon, C., Hong, Z. and Verma, D.P.S. (1994) Nodulin-24 follows a novel pathway for integration into the peribacteroid membrane in soybean root nodules. *J. Biol. Chem.* **269** (9), 6598–6602.

Curie, C., Alonso, J.M., Le Jean, M., Ecker, J.R. and Briat, J.F. (2000) Involvement of NRAMP1 from *Arabidopsis thaliana* in iron transport. *Biochem. J.* **347**, 749–755.

Day, D.A., Price, G.D. and Udvardi, M.K. (1989) Membrane interface of the *Bradyrhizobium japonicum–Glycine max* symbiosis: peribacteroid units from soybean nodules. *Aust. J. Plant Physiol.* **16**, 69–84.

Day, D.A., Kaiser, B.N., Thomson, R., Udvardi, M.K., Moreau, S. and Puppo, A. (2001) Nutrient transport across symbiotic membranes from legume nodules. *Aust. J. Plant Physiol.* **28**, 667–674.

Delves, A.C., Matthews, A., Day, D.A., Carter, A.S., Carroll, B.J. and Gresshoff, P.M. (1986) Regulation of the soybean – rhizobium nodule symbiosis by shoot and root factors. *Plant Physiol.* **82**, 588–590.

Dix, D.R., Bridgham, J.T., Broderius, M.A., Byersdorfer, C.A. and Eide, D.J. (1994) The *fet4* gene encodes the low-affinity Fe(II) transport protein of *Saccharomyces cerevisiae. J. Biol. Chem.* **269**, 26092–26099.

Dubois, E. and Grenson, M. (1979) Methylamine/ammonium uptake systems in *Saccharomyces cerevisiae*: multiplicity and regulation. *Mol. Gen. Genet.* **175**, 67–76.

Eide, D., Davis-Kaplan, S., Jordan, I., Sipe, D. and Kaplan, J. (1992) Regulation of iron uptake in *Saccharomyces cerevisiae. J. Biol. Chem.* **267**, 20774–20781.

Eide, D., Broderius, M., Fett, J. and Guerinot, M.L. (1996) A novel iron-regulated metal transporter from plants identified by functional expression in yeast. *Proc. Natl. Acad. Sci. USA*, **93**, 5624–5628.

Fleming, M.D., Trenor, C.C., Su, M.A., Foernzler, D., Beier, D.R., Dietrich, W.F. and Andrews, N.C. (1997) Microcytic anaemia mice have a mutation in *Nramp2*, a candidate iron transporter gene. *Nat. Genet.* **16**, 383–386.

Gietz, D., StJean, A., Woods, R.A. and Schiestl, R.H. (1992) Improved method for high-efficiency transformation of intact yeast cells. *Nucl. Acids Res.* **20**, 1425–1420.

Gonzalez, A., Koren'kov, V. and Wagner, G.J. (1999) A comparison of Zn, Mn, Cd and Ca-transport mechanisms in oat root tonoplast vesicles. *Physiologia Plantarum*, **106**, 203–209.

Grotz, N. and Guerinot, M.L. (2002) Limiting nutrients: an old problem with new solutions? *Curr. Opin. Plant Biol.* **5**, 158–163.

Gruenheid, S., Canonne-Hergaux, F., Gauthier, S., Hackam, D.J., Grinstein, S. and Gros, P. (1999) The iron-transport protein NRAMP2 is an integral membrane glycoprotein that colocalizes with transferrin in recycling endosomes. *J. Exp. Med.* **189**, 831–841.

Gunshin, H., Mackenzie, B., Berger, U.V., Gunshin, Y., Romero, M.F., Boron, W.F., Nussberger, S., Gollan, J.L. and Hediger, M.A. (1997) Cloning and characterization of a mammalian proton-coupled metal ion transporter. *Nature*, **388**, 482–488.

Hurkman, W.J. and Tanaka, C.K. (1986) Solubilization of plant-membrane proteins for analysis by two-dimensional gel electrophoresis. *Plant Physiol.* **81**, 802–806.

Kaiser, B.N., Finnegan, P.M., Tyerman, S.D., Whitehead, L.F., Bergersen, F.J., Day, D.A. and Udvardi, M.K. (1998) Characterization of an ammonium transport protein from the peribacteroid membrane of soybean nodules. *Science*, **281**, 1202–1206.

Kyte, J. and Doolittle, R.F. (1982) A simple method for displaying the hydropathic character of a protein. *J. Mol. Biol.* **157**, 105–132.

Laemmli, U.K. (1970) Cleavage of structural proteins during the assembly of the head of bacteriophage T4. *Nature*, **227**, 680–685.

Levier, K., Day, D.A. and Guerinot, M.L. (1996) Iron uptake by symbiosomes from soybean root nodules. *Plant Physiol.* **111**, 893–900.

Levier, K. and Guerinot, M.L. (1996) The *Bradyrhizobium japonicum fega* gene encodes an iron-regulated outer membrane protein with similarity to hydroxamate-type siderophore receptors. *J. Bacteriol.* **178**, 7265–7275.

Mellor, R.B. (1989) Bacteroids in the *Rhizobium*-legume symbiosis inhabit a plant internal lytic compartment: implications for other microbial endosymbioses. *J. Exp. Bot.* **40**, 831–839.

Minet, M., Dufour, M. and Lacroute, F. (1992) Complementation of *Saccharomyces cerevisiae* auxotrophix mutants by *Arabidopsis thaliana* cDNAs. *Plant J.* **2**, 417–422.

Moreau, S., Meyer, J.M. and Puppo, A. (1995) Uptake of iron by symbiosomes and bacteroids from soybean nodules. *FEBS Lett.* **361**, 225–228.

Moreau, S., Day, D.A. and Puppo, A. (1998) Ferrous iron is transported across the peribacteroid membrane of soybean nodules. *Planta*, **207**, 83–87.

Moreau, S., Thomson, R.M., Kaiser, B.N., Trevaskis, B., Guerinot, M.L., Udvardi, M.K., Puppo, A. and Day, D.A. (2002) GmZIP1 encodes a symbiosis-specific zinc transporter in soybean. *J. Biol. Chem.* **277**, 4738–4746.

Peterson, G.L. (1977) A simplification of the protein assay of Lowry et al. which is more generally applicable. *Anal. Biochem.* **83**, 346–356.

Rodrigues, V., Cheah, P.Y., Ray, K. and Chia, W. (1995) Malvolio, the *Drosophila* homologue of mouse Nramp-1 (bcg), is expressed in macrophages and in the nervous system and is required for normal taste behaviour. *EMBO J.* **14**, 3007–3020.

Romheld, V. (1987) Different strategies for iron acquisition in higher plants. *Physiol. Plant.* **70**, 231–234.

Sambrook, J., Fritsch, E.F. and Maniatis, T. (1989) *Molecular Cloning: a Laboratory Manual.* Cold Spring Harbour; Cold Spring Harbour Laboratory Press.

Supek, F., Supekova, L., Nelson, H. and Nelson, N. (1996) A yeast manganese transporter related to the macrophage protein involved in conferring resistance to mycobacteria. *Proc. Natl. Acad. Sci. USA*, **93**, 5105–5110.

Tabuchi, M., Yoshimori, T., Yamaguchi, K., Yoshida, T. and Kishi, F. (2000) Human NRAMP/DMT1, which mediates iron transport across endosomal membranes, is localised to late endosomes and lysosomes in HEp-2 cells. *J. Biol. Chem.* **275**, 22220–22228.

Tabuchi, M., Tanaka, N., Nishida-Kitayama, J., Ohno, H. and Kishi, F. (2002) Alternative splicing regulates the subcellular localisation of divalent metal transporter 1 isoforms. *Mol. Biol. Cell*, **13**, 4371–4387.

Tang, C., Robson, A.D. and Dilworth, M.J. (1990) A split-root experiment shows that iron is required for nodule initiation in *Lupinus angustifolius* L. *New Phytol.* **115**, 61–67.

Thomine, S., Wang, R.C., Ward, J.M., Crawford, N.M. and Schroeder, J.I. (2000) Cadmium and iron transport by members of a plant metal transporter family in *Arabidopsis* with homology to Nramp genes. *Proc. Natl. Acad. Sci. USA*, **97**, 4991–4996.

Udvardi, M.K. and Day, D.A. (1997) Metabolite transport across symbiotic membranes of legume nodules. *Annu. Rev. Plant Physiol. Plant Mol. Biol.* **48**, 493–523.

Vert, G., Grotz, N., Dedaldechamp, F., Gaymard, F., Guerinot, M.L., Briata, J.F. and Curie, C. (2002) IRT1, an *Arabidopsis* transporter essential for iron uptake from the soil and for plant growth. *Plant Cell*, **14**, 1223–1233.

Vidal, S.M. and Gros, P. (1994) Resistance to infection with intracellular parasites – identification of a candidate gene. *News Physiol. Sci.* **9**, 178–183.

Whitehead, L.F. and Day, D.A. (1997) The peribacteroid membrane. *Physiologia Plantarum*, **100**, 30–44.

Wittenberg, J.B., Wittenberg, B.A., Day, D.A., Udvardi, M.K. and Appleby, C.A. (1996) Siderophore-bound iron in the peribacteroid space of soybean root nodules. *Plant Soil*, **178**, 161–169.

CHAPTER 19

Provided example article 2: Britton-Simmons and Abbott (2008)

Journal of Ecology 2008, **96**, 68–77

doi: 10.1111/j.1365-2745.2007.01319.x

Short- and long-term effects of disturbance and propagule pressure on a biological invasion

Kevin H. Britton-Simmons* and Karen C. Abbott†

Department of Ecology and Evolution, The University of Chicago, 1101 East 57th Street, Chicago, IL 60637, USA

Summary

1. Invading species typically need to overcome multiple limiting factors simultaneously in order to become established, and understanding how such factors interact to regulate the invasion process remains a major challenge in ecology.

2. We used the invasion of marine algal communities by the seaweed *Sargassum muticum* as a study system to experimentally investigate the independent and interactive effects of disturbance and propagule pressure in the short term. Based on our experimental results, we parameterized an integrodifference equation model, which we used to examine how disturbances created by different benthic herbivores influence the longer term invasion success of *S. muticum*.

3. Our experimental results demonstrate that in this system neither disturbance nor propagule input alone was sufficient to maximize invasion success. Rather, the interaction between these processes was critical for understanding how the *S. muticum* invasion is regulated in the short term.

4. The model showed that both the size and spatial arrangement of herbivore disturbances had a major impact on how disturbance facilitated the invasion, by jointly determining how much space-limitation was alleviated and how readily disturbed areas could be reached by dispersing propagules.

5. *Synthesis.* Both the short-term experiment and the long-term model show that *S. muticum* invasion success is co-regulated by disturbance and propagule pressure. Our results underscore the importance of considering interactive effects when making predictions about invasion success.

Key-words: biological invasion, biotic resistance, disturbance, establishment probability, propagule pressure, *Sargassum muticum*

Introduction

Biological invasions are a global problem with substantial economic (Pimentel *et al.* 2005) and ecological (Mack *et al.* 2000) costs. Research on invasions has provided important insights into the establishment, spread and impact of non-native species. One key goal of invasion biology has been to identify the factors that determine whether an invasion will be successful (Williamson 1996). Accordingly, ecologists have identified several individual factors (e.g. disturbance and propagule pressure) that appear to exert strong controlling influences on the invasion process. However, understanding how these processes interact to regulate invasions remains a major challenge in ecology (D'Antonio *et al.* 2001; Lockwood *et al.* 2005; Von Holle & Simberloff 2005).

Propagule pressure is widely recognized as an important factor that influences invasion success (MacDonald *et al.* 1989; Simberloff 1989; Williamson 1996; Lonsdale 1999; Cassey *et al.* 2005). Previous studies suggest that the probability of a successful invasion increases with the number of propagules released (Panetta & Randall 1994; Williamson 1989; Grevstad 1999), with the number of introduction attempts (Veltman *et al.* 1996), with introduction rate (Drake *et al.* 2005), and with proximity to existing populations of invaders (Bossenbroek *et al.* 2001). Moreover, propagule pressure may influence invasion dynamics after establishment by affecting the capacity of non-native species to adapt to their new environment (Ahlroth *et al.* 2003; Travis *et al.* 2005). Despite its acknowledged importance, propagule pressure has rarely been manipulated experimentally and the interaction of propagule pressure with other processes that regulate invasion success is not well understood (D'Antonio *et al.* 2001; Lockwood *et al.* 2005).

*Correspondence and present address. Friday Harbor Laboratories, University of Washington, 620 University Road, Friday Harbor, WA 98250, USA. E-mail: aquaman@u.washington.edu
†Present address: Department of Zoology, University of Wisconsin, 430 Lincoln Drive, Madison, WI 53706, USA

Resource availability is a second key factor known to influence invasion success and processes that increase or decrease resource availability therefore have strong effects on invasions (Davis *et al.* 2000). Resource pre-emption by native species generates biotic resistance to invasion (Stachowicz *et al.* 1999; Naeem *et al.* 2000; Levine *et al.* 2004). Consequently, physical disturbance can facilitate invasions by reducing competition for limiting resources (Richardson & Bond 1991; Hobbs & Huenneke 1992; Kotanen 1997; Prieur-Richard & Lavorel 2000). In most communities disturbances occur via multiple mechanisms and the disturbances created by different agents vary in their intensity and frequency (D'Antonio *et al.* 1999). Recent empirical (Larson 2003; Hill *et al.* 2005) and theoretical (Higgins & Richardson 1998) studies suggest that not all types of disturbance have equivalent effects on the invasion process. Moreover, most of what we know about the effects of disturbance on invasions comes from short-term experimental studies. It is presently unclear how different disturbance agents influence long-term patterns of invasion.

In order for any invasion to be successful, propagule arrival must coincide with the availability of resources needed by the invading species (Davis *et al.* 2000). Therefore, the interaction between propagule pressure and processes that influence resource availability will ultimately determine invasion success (Brown & Peet 2003; Lockwood *et al.* 2005; Buckley *et al.* 2007). In this study we used the invasion of shallow, subtidal kelp communities in Washington State by the Japanese seaweed *Sargassum muticum* as a study system to better understand the effects of propagule pressure and disturbance on invasion. In a factorial field experiment we manipulated both propagule pressure and disturbance in order to examine how these factors independently and interactively influence *S. muticum* establishment in the short term. We supplement the experimental results with a parameterized integrodifference equation model, which we use to examine how different natural disturbance agents influence the spread of *S. muticum* through the habitat in the longer term. Although a successful invasion clearly requires both establishment and spread of the invader, most studies have looked at just one of these processes (Melbourne *et al.* 2007). We take an integrative approach by employing both a short-term experiment and a longer-term model, allowing us to examine the effects of disturbance and propagule limitation on the entire invasion process.

Methods

STUDY SYSTEM

Our field research was based out of Friday Harbor Laboratories on San Juan Island, Washington State, USA. The field experiment was carried out at a site within the San Juan Islands Marine Preserve network adjacent to Shaw Island, known locally as Point George (48.5549° N, 122.9810° W). Field work was accomplished using SCUBA in shallow subtidal communities.

The native algal community characteristic of sheltered, rocky subtidal habitats in this region is species-rich and structurally complex (see Britton-Simmons 2006 for a more detailed description). In this ecosystem, space is an important limiting resource and in the

absence of disturbance there is little or no bare rock available for newly arriving organisms to colonize. This habitat has a diverse fauna of benthic herbivores, including molluscs and sea urchins, that create disturbances by clearing algae from the rocky substrata. The green sea urchin *Strongylocentrotus droebachiensis* is a generalist herbivore that reduces the abundance of native algae and creates relatively large disturbed patches (Vadas 1968; Duggins 1980). In the shallow zone where *S. muticum* is found, the green urchin is highly mobile and often occurs in aggregations (Paine & Vadas 1969; Foreman 1977; Duggins 1983; personal observation). Green urchins avoid areas where *S. muticum* is present because it is not a preferred food resource (Britton-Simmons 2004), but they can be found feeding in uninvaded areas adjacent to existing *S. muticum* populations (personal observation). Green urchins therefore create intermittent but relatively intense disturbances in areas where *S. muticum* is absent and some proportion of these disturbances can potentially be exploited by dispersing *S. muticum* propagules. In contrast, herbivorous benthic molluscs (chitons, limpets and snails) are ubiquitous in the shallow subtidal and unlike sea urchins they are unaffected by the presence of *S. muticum* (Britton-Simmons 2004). Herbivory by individual molluscs creates relatively small-scale disturbances, thereby providing a consistent supply of microsites that can be colonized by newly arriving species, including *Sargassum muticum* (see Appendix S1 in Supplementary Material for more information about mollusc diets).

THE INVADER

Sargassum muticum is a brown alga in the order Fucales that was introduced to Washington State in the early 20th century, probably with shipments of Japanese oysters that were imported for aquaculture beginning in 1902 (Scagel 1956). It is now common in shallow subtidal habitats throughout Puget Sound and the San Juan Islands (Nearshore Habitat Program 2001, personal observation). In the San Juan Islands, *S. muticum* has a pseudoperennial life history. Each holdfast produces as many as 18 laterals in the early spring, each of which can grow as tall as three metres. In late summer to early autumn the laterals senesce and are lost, leaving only the basal holdfast portion of the thallus to overwinter.

Sargassum muticum has a diplontic (uniphasic) life cycle, is monecious, and is capable of selfing. Reproduction typically occurs between late June and late August in our region. During reproduction the eggs of *S. muticum* are released from and subsequently adhere to the outside of small reproductive structures called receptacles. Once fertilized, the resulting embryos remain attached while they develop into tiny germlings (< 200 µm in length) with adhesive rhizoids (Deysher & Norton 1982). Germlings then detach from the receptacle and sink relatively quickly, recruiting in close proximity to the parent plant (Deysher & Norton 1982). Although most recruitment occurs within 5 m of adult plants, recruits have been found as far as 30 m from the nearest adult (Deysher & Norton 1982). Longer distance dispersal probably occurs when plants get detached from the substratum and subsequently become fertile after drifting for some period of time (Deysher & Norton 1982). One distinctive feature of the *S. muticum* invasion is that it is extremely limited in vertical extent. In the San Juan Islands, *S. muticum* is found from the low intertidal to the shallow subtidal zone (Norton 1977; personal observation), from approximately –0.5 m Mean Lower Low Water (MLLW) to –7 m MLLW. However, it is most abundant in the shallow subtidal, from approximately –2 m MLLW to –4 m MLLW. Thus, in areas where *S. muticum* has invaded it forms a narrow band along the shore.

FIELD EXPERIMENT

We used a two-way factorial design manipulating propagule pressure (six levels) and disturbance (two levels) with three replicates per treatment combination. Subtidal plots (30 cm × 30 cm) at a depth of 3–4 m below MLLW were selected so that differences in the identity and abundance of taxa, aspect, and relief were minimized and the plots were randomly assigned to treatments. None of the experimental plots contained *S. muticum* prior to the experiment. However, some *S. muticum* was present at Point George and it was removed prior to the reproductive season in order to prevent contamination of the experimental plots from external sources of *S. muticum* propagules.

The disturbance treatment had two levels: control and disturbed. Control plots were not altered in any way, but they did vary somewhat in how much natural disturbance had occurred in them prior to the experiment (mean = 7.7% of plot area). Plots in the disturbance treatment were scraped down to bare rock so that no visible organisms remained. These two treatments represent extremes in the levels of disturbance that are likely to occur in nature. The unaltered control plots contained a rich assemblage of native species. The disturbed plots were similar in spatial scale to a patch that a small group of urchins might create, but represent an unusually intense disturbance because all native species, including crustose coralline algae (which cover an average of 27.7% of the substratum at this depth), were removed. These treatments maximized our ability to detect an effect of disturbance in our experiment.

Immediately following the imposition of the disturbance treatment (July 2002) the plots were experimentally invaded by suspending 'brooding' *S. muticum* over them. This was accomplished by collecting *S. muticum* from the field and transporting them to the lab where the appropriate ratio of sterile to reproductive tissue (see below) was placed in 30 cm × 30 cm vexar bags. The bags were returned to the field the same day and suspended over the experimental plots for 1 week. Propagule pressure was manipulated by varying the ratio of sterile to reproductive tissue in the bags while holding the total biomass of *S. muticum* tissue constant. The propagule pressure treatment had six levels, corresponding to the following amounts of reproductive tissue (in grams): 0, 50, 100, 175, 250 and 350 (average mass of mature *S. muticum* in this region is 174 g). Based on propagule production–mass relationships derived by Norton & Deysher (1988) for *S. muticum*, we estimate that approximately 5 million propagules were released in each replicate of our highest propagule pressure treatment. We assumed a linear relationship between the mass of adult reproductive tissue and propagule output because we know of no *Sargassum* study that suggests otherwise. Sterile tissue was added to bags as necessary in order to bring the total biomass to 350 g. Reproductive and sterile tissue was mixed in the bags so that the reproductive tissue was well distributed throughout. This experimental manipulation mimics the level of propagule input that would occur in an incipient invasion or if a drifting plant became tangled with attached algae and subsequently released its propagules.

Recruitment of *S. muticum* was quantified by counting the number of *S. muticum* juveniles that were present in the plots 5 months after the experimental invasion, which is the earliest they can reliably be seen in the field. We resurveyed the plots to count the number of *S. muticum* adults present 11 months after the invasion (just prior to reproductive season) and then removed all *S. muticum* from the experimental plots in order to prevent it from spreading.

STATISTICAL ANALYSIS

We analysed the *S. muticum* recruitment data using a two-way ANOVA followed by separate regression analyses on each disturbance treatment. For the control treatment, we performed a multiple regression to determine what proportion of recruitment variation was explained by propagule input and space availability. For the disturbed plots, which did not vary in the amount of available space, we carried out a simple linear regression to determine the impact of propagule input on recruitment. We used the results of these analyses to inform the construction of mechanistic candidate functions for the relationship between propagule input, space availability and recruitment. These candidate functions were compared using differences in the Akaike's information criteria (AIC differences; Burnham & Anderson 2002). We then used model averaging, a form of multimodel inference in which parameter estimates from more than one candidate function are used jointly to describe the data, in order to select a parameterized recruitment function for the *S. muticum* spread model.

The *S. muticum* survivorship data did not conform to the assumptions of ANOVA (even after a number of different transformations) so we used a non-parametric Kruskal–Wallis test to ask whether *S. muticum* survivorship differed in the disturbed and control treatments. We then fitted five different survivorship functions, assuming binomial error, to the data to test whether *S. muticum* survivorship (number of adults per recruit) was density-dependent. Because the Kruskal–Wallis test suggested that survivorship differed significantly between the two disturbance treatments (see Results) we chose to fit the models to those two treatments separately to test for density dependence. In addition to type 1 (linear), type 2 (saturating), and type 3 (sigmoidal) functions, we also fitted a constant survivorship model. These candidate functions were compared using the Akaike's information criterion (AIC differences; Burnham & Anderson 2002).

The numbers of adult *S. muticum* (after 11 months) also violated the assumptions of ANOVA (despite transformations), so we used non-parametric statistics to test two hypotheses: (i) adult density is independent of disturbance treatment (Wilcoxon Signed Ranks Test), and (ii) adult density is independent of propagule pressure treatment (Kruskal–Wallis Test).

MODEL

We used an integrodifference equation (IDE) model to describe the spatial spread of an *S. muticum* population. IDE models assume that the habitat is continuous in space, and that reproduction and dispersal occur in discrete bouts. The depths inhabited by *S. muticum* comprise a relatively narrow vertical band, so the spread of the population was assumed to occur in a one-dimensional habitat. The model follows two state variables through time. $N_t(x)$ is the density of *S. muticum* at a location x along this habitat at time t, and $Z_t(x)$ is the amount of bare rock at x during t. The values for these state variables are determined by functions representing the important ecological processes in this system. *Sargassum muticum* density is determined by the production and recruitment of propagules and by adult survival. Bare rock is created by benthic herbivore disturbances, since herbivores consume native algae and thus alleviate space limitation. The form of our model is then

$$N_{t+1}(x) = sP_t(x)f(P_t(x), Z_t(x)) + rN_t(x), \qquad \text{eqn 1}$$

$$Z_{t+1}(x) = (1 - \eta_t(x))gZ_t(x) + \eta_t(x)A. \qquad \text{eqn 2}$$

$P_t(x)$ is the number of propagules at location x at the start of year t, and equals the number of propagules produced at x and remaining near their parent plant plus the sum of propagules from all other locations within the habitat (with endpoints a and b) which disperse to x. $P_t(x)$ is governed by the equation $P_t(x) = \int_a^b \omega N_t(y)k(x-y)dy$.

Each adult produces ω propagules and their dispersal is described by the function k. The function $f(P_t(x), Z_t(x))$ in equation 1 gives the fraction of propagules which successfully recruit, given that the amount of bare rock at location x equals $Z_t(x)$ and there is an initial input of $P_t(x)$ propagules. Based on data from the experiment, we assume that recruitment function has the form $f(P_t(x), Z_t(x)) = \rho_1(Z_t(x) + \rho_2)^{\rho_5} P_t(x)/[1 + \rho_3(Z_t(x) + \rho_2)^{\rho_5} + \rho_4 P_t(x)^2]$, with values for the ρ_i and methods for fitting this function given in Appendix S2. s and r are fractions of germlings and adults, respectively, that survive to the following year. Parameters for *Sargassum* fecundity and dispersal were attained from the literature (Deysher & Norton 1982; Norton & Deysher 1988) and all other parameter values used in our simulations were estimated from our own field data. The methods and results for fitting parameters are given in Appendix S2.

In equation 2, $\eta_t(x)$ is the proportion of the habitat scraped clear by grazers. If left ungrazed, we assumed that bare rock at a given location experiences geometric decay, with rate g, as it becomes utilized by native algae. The parameter A in equation 2 is a scaling constant representing the size of the habitable area at each point x. We modelled benthic herbivore disturbance in two different ways. First, we constructed a stochastic model for $\eta_t(x)$ based on our understanding of the natural history of the system. Second, we built a more generalized stochastic model for $\eta_t(x)$. In the *S. muticum* system, bare rock is generated in small patches when an area is grazed by molluscs (chitons and limpets), or in larger patches by sea urchin grazing. Both types of disturbance create bare rock for *S. muticum* to potentially exploit, and the disturbance types differ only in their size and spatial distribution. We assumed that the mollusc disturbances are ubiquitous, whereas large urchin-grazed areas are patchily distributed across the habitat. Due to uncertainty in the exact size and frequency of these disturbances, we ran simulations over a very wide range of possible parameter values. In the generalized model for $\eta_t(x)$, we allowed disturbances of any size to occur with any degree of spatial aggregation, rather than requiring large disturbances to be patchy and small ones to be spread throughout the habitat. Our methods for drawing values for $\eta_t(x)$ in these simulations are described in Appendix S3 and summarized in Table C.1 therein.

In our system, native benthic grazers do not eat *S. muticum* adults (Britton-Simmons 2004; personal observation), but it is unknown whether they will consume new *S. muticum* recruits when they are very small (e.g. Sjøtun *et al.* 2007) and hence difficult to avoid ingesting incidentally. Whether or not disturbance events can directly cause mortality of the invader can be very important in determining invasion success (Buckley *et al.* 2007). In our simulations, we therefore considered both the case where *S. muticum* is never eaten by grazers, and the case where *S. muticum* is eaten at the rate $\eta_t(x)$ until it reaches the age of 1 year.

Results

The field experiment showed that recruitment of *S. muticum* was higher in plots that were disturbed compared to control plots (Fig. 1a) suggesting that resource availability limited recruitment. Increasing propagule pressure led to significant increases in average *S. muticum* recruitment in both distur-

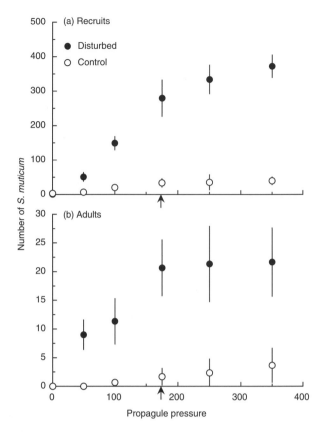

Fig. 1. Number of *Sargassum muticum* (a) recruits and (b) adults in field experiment plots (900 cm²). Propagule pressure is grams of reproductive tissue suspended over experimental plots at beginning of experiment. The average mass of an adult *S. muticum* (174 g) is indicated by an arrow. Data are means ± 1 SE ($n = 3$).

bance treatments (Fig. 1a). Finally, a significant interaction between disturbance and propagule pressure ($F_{5,24} = 3.77$, $P = 0.01$) indicates that the plots in the two disturbance treatments differed in the extent to which they were limited by propagule availability. Multiple regression analysis of the *S. muticum* recruitment data from the control treatment, with space and propagule input as continuous explanatory variables, explained most of the recruitment variability ($R^2 = 0.87$, Fig. 1a). This analysis showed that both space (Fig. 1a, $b = 0.703$, $P < 10^{-4}$) and propagule treatment (Fig. 1a, $b = 0.657$, $P < 10^{-3}$) had strong influences on recruitment in the control treatment. Because there was no variation in space availability in the disturbed treatment, we used simple linear regression analysis to examine the relationship between propagule input and *S. muticum* recruitment in the disturbed treatment (Fig. 1a, $R^2 = 0.84$, $P < 10^{-6}$). The results suggest that in the absence of space limitation propagule input explains most of the variability in *S. muticum* recruitment.

We used these results to create a set of mechanistic candidate functions for the relationship between *S. muticum* recruitment, propagule pressure and space availability (see Appendix S2). The only candidate models supported by the data (AIC differences < 4; Burnham & Anderson 2002) show a type 3 (sigmoidal) relationship between propagule pressure and recruitment, and either a type 2 (saturating) or type 3 relationship between available space and recruitment (Appendix S2, Table B.1). Due to practical constraints on the number of treatments that could be replicated in the field, we have data only on very low available space (control plots) and very high available space (disturbed plots), and insufficient data at intermediate values to resolve the functional relationship between space-limitation and recruitment. We therefore used model averaging (Burnham & Anderson 2002) to combine our parameter estimates for the two supported models and used the resulting function to describe space- and propagule-limitation in recruitment in the simulation model. We also ran simulations using each of the supported recruitment models separately. The results from the two supported models and the averaged model were very similar, so we present results only from the averaged model.

Survivorship (from 5 months to 11 months of age) of *S. muticum* was significantly higher in disturbed plots ($U = 76.5$, $P < 0.05$). Mean survivorship (± 1 SD) in control plots was 3.4% ($\pm 3.8\%$), compared to 6.1% ($\pm 2.2\%$) in disturbed plots. Our analysis of survivorship as a function of recruitment density suggests density-independence (Appendix S2, Table B.2), so we used the mean survivorship across all experimental plots as the germling survival rate (s) in our model.

Simulations of the parameterized model under various disturbance regimes reveal several interesting patterns. Using the disturbance scenario with ubiquitous mollusc disturbances and large, patchily distributed urchin disturbances, we found that a single adult *S. muticum* was almost always sufficient to start a successful invasion. This is in agreement with our empirical observation that propagule input always resulted in positive recruitment, even in space-poor control plots. We quantified population growth in our model by reporting the density of *S. muticum* after 100 years, averaged across the invaded area, and we use the length of habitat occupied by *S. muticum* after 100 years as a measure of invasion rate. When we assumed that *S. muticum* was never consumed by benthic herbivores, both the mean *S. muticum* population density and the length of the invaded area increased with both the mean intensity of mollusc grazing and with the size and number of urchin disturbances (Fig. 2, solid lines). Changing the variance in the intensity of mollusc grazing had essentially no effect (not shown). Unless urchin disturbances were extremely large and numerous (top 3 lines, Fig. 2g–j), the mollusc grazing had a much stronger effect on *S. muticum* density than did urchin grazing.

When we assumed that native grazers eat *S. muticum* germlings, *S. muticum* density and the length of habitat invaded still increased with the intensity of mollusc disturbance, as long as molluscs grazed less than 50% of the habitat bare (Fig. 2, dashed lines). Actual mollusc disturbances are typically much smaller than 50% (personal observation). Indeed, we note that if all of the bare rock in the experiment's control plots was attributed to mollusc grazing, the average grazing intensity would be only 7.7%. Within the realistic range of parameter values, then, molluscs facilitate the invasion in the model even when they consume young *S. muticum*.

Urchin disturbances that were few and/or small had little effect on the invasion, but large and numerous urchin disturbances decreased the final *S. muticum* density and the size of the invaded area when grazers consumed new recruits (Fig. 2e–j). *Sargassum muticum* failed to establish when urchin disturbances were both very large (20–50 m of linear habitat scraped bare per disturbance) and extremely abundant (100–200 such disturbances per year). These results are corroborated by the generalized model of disturbance, which showed that when the total proportion of the habitat disturbed per year is held constant smaller disturbances affecting a greater number of locations resulted in the highest final *S. muticum* densities and invaded areas (Appendix S2, Fig. C.1). When these disturbed locations were more clumped in space, this resulted in a slight decrease in the final size of the invaded area.

The treatment effects were still apparent when adults were counted at the end of the experiment (Fig. 1b). Adult *S. muticum* density was higher in the disturbed treatment than in the control treatment ($Z = -3.41$, $P < 0.001$). In addition, adult *S. muticum* density appeared to be positively related to propagule pressure (Fig. 1b, $H_5 = 16.10$, $P = 0.006$), with high propagule pressure resulting in a maximum of between 20 and 25 adults per plot (900 cm^2).

How was the probability of successful invasion influenced by propagule pressure? We defined successful invasion of an experimental plot as the presence of one or more adult *S. muticum* at the end of the experiment (11 months after invasion). We consider this a reasonable way to define invasion success given that reproduction of these adults was imminent (< 1 month away), survivorship is very high at this life-history stage (Appendix S2, Table B.3), and both our model and experimental results indicate that a single individual is capable of establishing a population. We plotted the proportion of plots in each treatment combination that were successfully invaded as a function of propagule pressure (Fig. 3). Because we had only three replicates per treatment combination the probability values were constrained to four possible values (0, 0.33, 0.66, or 1.0). In addition, we tested only six levels of propagule input and therefore have limited capacity to resolve the details of this relationship. Therefore, we did not attempt to fit statistical models to these data. In disturbed plots, invasion was certain even at the lowest level of propagule pressure in our experiment (Fig. 3). However, in control plots the probability of invasion was less than 1 until propagule pressure reached a level of 250 g of reproductive tissue, an amount of tissue greater than the average mass of an adult *S. muticum* (Fig. 3).

Discussion

Our experimental results demonstrate that space- and propagule-limitation both regulate *S. muticum* recruitment. Our finding that *S. muticum* recruitment was positively related to propagule input is similar to those of two previous studies (Parker 2001; Thomsen *et al.* 2006), in which the propagule input of invasive plants was manipulated. In our control

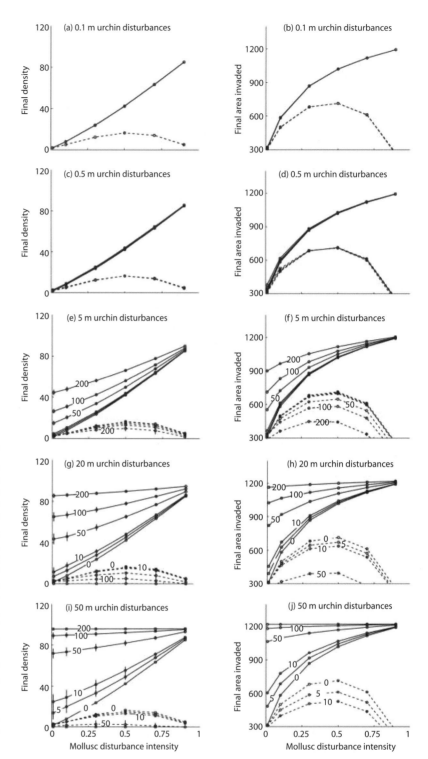

Fig. 2. Simulation results using the mollusc/urchin model for disturbance. The first column (a, c, e, g and i) shows the mean *Sargassum muticum* density (individuals per 900 cm²) and the second column (b, d, f, h and j) show the length of habitat occupied (metres) after 100 years. Solid lines are the results when native grazers never eat *S. muticum* and dashed lines are results when *S. muticum* recruits (less than 1 year old) are eaten by grazers. The *x*-axis in all plots shows the average proportion of rock scraped bare by molluscs. The number superimposed on each line is the number of urchin disturbances per year (numbers are omitted when the lines overlap completely or are very close together). The mean size of these urchin disturbances increases from the top row (a–b) to the bottom (i–j) and is printed at the top of each graph. Error bars, when large enough to be visible, are ± 1 SE (*n* = 100, as averages were taken across two values for the variance in mollusc intensity with 50 replicates each).

treatment space was limiting, a result that has also been found in previous studies of *S. muticum* recruitment (Deysher & Norton 1982; De Wreede 1983; Sanchez & Fernandez 2006). Consequently, increasing propagule pressure had a relatively weak effect on recruitment in undisturbed plots (Fig. 1a). However, when space limitation was alleviated by disturbing the plots, increasing propagule pressure caused a dramatic increase in recruitment (Fig. 1a). This suggests that in the presence of adequate substratum for settlement, propagule

limitation becomes the primary factor controlling *S. muticum* recruitment. These results indicate that *S. muticum* recruitment under natural field conditions will be determined by the interaction between disturbance and propagule input.

Only a few previous studies have investigated the effect of resource supply on the relationship between propagule pressure and recruitment of an introduced species. Although disturbance generally increases invasion success by increasing resource availability (Richardson & Bond 1991; Bergelson

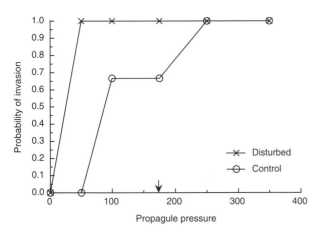

Fig. 3. Probability of invasion as a function of propagule pressure. Probability of invasion is the proportion of plots in each treatment combination ($n = 3$) that contained at least one adult *Sargassum muticum* at the end of the experiment. The average mass of an adult *S. muticum* (174 g) is indicated by an arrow.

et al. 1993; Levin *et al.* 2002; Valentine & Johnson 2003; Clark & Johnston 2005), Parker (2001) found evidence that disturbance reduced Scotch broom (*Cystisus scoparius*) recruitment from seed at all levels of propagule input. This effect occurred because the native flora actually facilitated Scotch broom germination, probably by increasing soil moisture and/or nutrients (Parker 2001). Similarly, Thomsen *et al.* (2006) showed that in the absence of a water addition treatment establishment of an exotic perennial grass was greatly reduced, even at high levels of propagule input. Finally, Valentine & Johnson (2003) found that disturbance facilitated invasion by the introduced kelp *Undaria pinnatifida* even when propagule pressure was high. These studies and our own work provide empirical evidence that the interaction between propagule input and the biotic and abiotic processes that mediate resource availability will be key to understanding patterns of invasion.

The effects of the disturbance and propagule pressure treatments that were manifest in the *S. muticum* recruitment data persisted until the end of the experiment (Fig. 1b). That adult *S. muticum* density was higher in the disturbed treatment than in the control treatment suggests that disturbance may increase the population growth rate of *S. muticum* during the initial stages of the invasion. Natural disturbances that are less intense than our experimental scrapings might have a more modest effect on *S. muticum* density, but our simulation results suggest that even small disturbances can play a major role in facilitating the invasion. Our simulations further suggest that this effect should persist over long time-scales (Fig. 2).

In subtidal habitats both biotic and abiotic disturbances occur, but it is doubtful that they are both relevant to the *S. muticum* invasion in this system. Consumption of algae by the diverse fauna of benthic herbivores in this system (see Methods) is a common and consistent source of disturbance that is likely to be relevant to the *S. muticum* invasion and was therefore the focus of our model. Abiotic disturbances are unlikely to play an important role in this regard because tidal currents are not a substantial cause of algal mortality in this region (Duggins *et al.* 2003) and the inland waters of Puget Sound, the San Juan Islands and the Strait of Georgia are protected from the ocean swells that play a key role on the outer coast of Washington State. Although locally generated storm waves are an important source of disturbance during the winter (Duggins *et al.* 2003), storms during the summer months when *S. muticum* is reproductive are rare.

SIMULATED URCHIN/MOLLUSC DISTURBANCES

In addition to enhancing *S. muticum* recruitment, disturbance increased the survivorship of juvenile *S. muticum*. In our system, the green urchin (*Strongylocentrotus droebachiensis*) creates relatively large disturbed patches and *S. muticum* that recruit to these patches probably benefit from reduced competition with native algae. Unlike other systems where sea urchins feed on both native and non-native algae alike (Valentine & Johnson 2005), green urchins do not consume adult *S. muticum* (Britton-Simmons 2004) although it is possible that they incidentally consume new recruits. Studies in other systems have also reported positive effects of disturbance on the survivorship of non-native species (Gentle & Duggin 1997; Williamson & Harrison 2002). In general, disturbance probably enhances survivorship because it reduces the size or abundance of native species that compete for resources with invaders (Gentle & Duggin 1997; Britton-Simmons 2006). Indeed, our modelling results suggest that even when juvenile survivorship is reduced by herbivory, the net effect of grazers is still usually positive (Fig. 2).

The simulation model suggested that not all disturbance agents have equivalent effects on space-limitation. Small bare patches throughout the habitat facilitated *S. muticum* spread (Fig. 2 and Appendix S3, Fig. C.1) by increasing the amount of bare rock near any given reproductive adult. Molluscs are ubiquitous in these subtidal habitats and although they typically create very small disturbances, the model suggests that this is sufficient for *S. muticum* to successfully invade, even in the absence of other disturbance agents (e.g. urchins and humans).

Urchins create much larger open spaces, but urchin disturbances could not be used by settling propagules unless a reproductive adult happened to be nearby or a long-distance dispersal event occurred. When there are many urchin disturbances in a year, the chance that such a disturbance occurs near an *S. muticum* adult increases and, because long-distance propagule dispersal is rare, this greatly enhances the likelihood that a propagule will reach the disturbed area. Accordingly, small numbers of urchin disturbances in our model did not affect the spread of *S. muticum* (Fig. 2a–d), but numerous and sufficiently large disturbances did (Fig. 2e–j). Washington State is at the southern end of the green urchin's range in the eastern Pacific and at the majority of sites in the San Juan Islands this species is absent or at relatively low abundance. Consequently, molluscs are probably the most important source of disturbance for *S. muticum* in this region; green urchins may be a more important disturbance agent in more northerly portions of its range (where it reaches higher densities). That urchin disturbance was not necessary for successful invasion by *S. muticum* in the model is an important result because *S. muticum* has invaded many areas in this region where urchins are absent. Indeed, urchins avoid areas where *S. muticum* is present (Britton-Simmons 2004) and since this effect was not included in the model, urchin disturbances probably contribute even less to *S. muticum* spread than our simulations suggest.

PROPAGULE PRESSURE AND INVASION SUCCESS

How much invasion risk does a given level of propagule pressure pose? Previous studies have demonstrated a positive relationship between propagule pressure and the establishment success of non-native species (Grevstad 1999; Parker 2001; Ahlroth *et al.* 2003; Cassey *et al.* 2005). However, we know very little about the relationship between establishment probability and propagule pressure or the factors that affect it (Lockwood *et al.* 2005). Possibilities include a linear relationship (Lockwood *et al.* 2005) as well as more complex relationships containing thresholds or other non-linearities (Griffith *et al.* 1989; Ruiz & Carlton 2003; Lockwood *et al.* 2005; Buckley *et al.* 2007). Our experimental results suggest that the relationship is non-linear (Fig. 3). Indeed, all communities in which abiotic factors do not preclude invasion are probably vulnerable to invasion such that above some threshold level of propagule input successful invasion is a virtual certainty. Consequently, this relationship must be nonlinear because by definition it saturates at a probability of one. In our system disturbance appeared to reduce the level of propagule pressure necessary to ensure invasion success. However, even control plots had a high probability of invasion once the level of propagule pressure exceeded that produced by an average adult *S. muticum*. Unfortunately, the limited number of treatment levels in our experiment constrains our ability to resolve the details of this relationship. Nevertheless, in the control treatment there was some evidence of a threshold level of propagule pressure below which invasion was very unlikely to occur (Fig. 3).

Our model reflects what we believe to be the most important factors limiting invasion success (propagule-limitation and competition for space) but other factors we did not include in the model, such as stochastic mortality, density-dependent mortality of adults, competition with native species for resources besides space (e.g. light, Britton-Simmons 2006) and abiotic conditions, could constrain *S. muticum*'s distribution and abundance in the field. Empirical studies have demonstrated the importance of biotic resistance in regulating invasions (see reviews by Levine & D'Antonio 1999; Levine *et al.* 2004) and the community that *S. muticum* is invading is no exception (Britton-Simmons 2006). However, some authors have suggested that propagule pressure has the potential to overcome biotic resistance (D'Antonio *et al.* 2001; Lockwood *et al.* 2005). Levine (2000) found that seed supply overpowered biotic resistance that was generated by plant communities at small spatial scales (18 cm × 18 cm). A more recent terrestrial experiment also reported that propagule pressure was the primary determinant of invasion success, overwhelming the effects of other factors, such as disturbance and resident diversity, which were concurrently manipulated (Von Holle & Simberloff 2005). However, 'propagules' in that study were seedlings transplanted into experimental plots and seedlings may not be regulated by the same factors as seeds, which are the life stage responsible for invasion spread in natural systems. Nevertheless, if propagule pressure can indeed overcome those factors that were not included in our model then one might ask why *S. muticum* has not completely taken over the shallow subtidal zone in this system, as our model predicts under most disturbance regimes. Interestingly, whether *S. muticum* is indeed in the process of doing so is not entirely clear. There are very few areas in the San Juan region where *S. muticum* is completely absent at the appropriate depths (personal observation), yet at many sites *S. muticum* is currently at low abundance and it is unclear whether these sites represent incipient invasions or whether something is inhibiting local population growth.

Conclusions

In our system, neither disturbance nor propagule input alone was sufficient to maximize invasion success (i.e. establishment probability and invader population density). Increasing propagule pressure had relatively little effect on total recruitment in control plots (Fig. 1a), though at high levels it ultimately overcame space limitation and ensured successful invasion (Fig. 3). However, even at high levels of propagule input, final *S. muticum* density was low in the absence of disturbance (Fig. 1b). Based on our experimental results alone, we might have predicted strong effects of both molluscs and urchins on the *S. muticum* invasion in the long term. However, the simulation model suggested that these two natural disturbance agents should have different effects on long-term invasion due to differences in the spatial structure of these disturbances. The model results demonstrate that caution should be exercised when extrapolating the results of short-term disturbance experiments over longer time intervals. In this marine community invasion success was co-regulated by propagule pressure and biotic resistance. Our results underscore the importance of considering interactive effects when making predictions about invasion success.

Acknowledgements

Thanks to Ben Pister, Sam Sublett and Jake Gregg for SCUBA assistance in the field. For helpful discussions that improved this work we thank Timothy Wootton, Cathy Pfister, Greg Dwyer, Joy Bergelson, Mathew Leibold, Spencer Hall and Bret Elderd. Yvonne Buckley, Barney Davies and an anonymous referee provided very helpful comments on an earlier version of the manuscript. The director and staff of Friday Harbor Laboratories provided logistical support and access to laboratory and SCUBA facilities. The field research was funded by a grant to Timothy Wootton from The SeaDoc Society at UC Davis, and both authors were supported by a Graduate Assistance in Areas of National Need Training Grant (P200A040070) during the completion of this work.

References

Ahlroth, P., Alatalo, R., Holopainen, A., Kumpulainen, T. & Suhonen, V. (2003) Founder population size and number of source populations enhance colonization success in waterstriders. *Oecologia*, **137**, 617–620.

Bergelson, J., Newman, J.A. & Floresroux, E.M. (1993) Rates of weed spread in spatially heterogeneous environments. *Ecology*, **74**, 999–1011.

Bossenbroek, J.M., Kraft, C.E. & Nekola, J.C. (2001) Prediction of long-distance dispersal using gravity-models: zebra mussel invasion of inland lakes. *Ecological Applications*, **11**, 1778–1788.

Britton-Simmons, K.H. (2004) Direct and indirect effects of the introduced alga *Sargassum muticum* on benthic, subtidal communities of Washington State, USA. *Marine Ecology Progress Series*, **277**, 61–78.

Britton-Simmons, K.H. (2006) Functional group diversity, resource preemption and the genesis of invasion resistance in a community of marine algae. *Oikos*, **113**, 395–401.

Brown, R.L. & Peet, R.K. (2003) Diversity and invasibility of southern Appalachian plant communities. *Ecology*, **84**, 32–39.

Buckley, Y.M., Bolker, B.M. & Rees, M. (2007) Disturbance, invasion, and re-invasion: managing the weed-shaped hole in disturbed ecosystems. *Ecology Letters*, **10**, 809–817.

Burnham, K.P. & Anderson, D.R. (2002) *Model Selection and Inference: A Practical Information–Theoretic Approach*. Springer Publishing, New York, NY.

Cassey, P., Blackburn, T.M., Duncan, R.P. & Lockwood, J.L. (2005) Lessons from the establishment of exotic species: a meta-analytical case study using birds. *Journal of Animal Ecology*, **74**, 250–258.

Clark, G.F. & Johnston, E.L. (2005) Manipulating larval supply in the field: a controlled study of marine invasibility. *Marine Ecology Progress Series*, **298**, 9–19.

D'Antonio, C.M., Dudley, T.L. & Mack, M. (1999) Disturbance and biological invasions: direct effects and feedbacks. *Ecosystems of the World 16: Ecosystems of Disturbed Ground* (ed. L.R. Walker), pp. 413–452. Elsevier.

D'Antonio, C.M., Levine, J. & Thomsen, V. (2001) Ecosytem resistance and the role of propagule supply: a California perspective. *Journal of Mediterranean Ecology*, **2**, 233–245.

Davis, M.A., Grime, J.P. & Thomson, K. (2000) Fluctuating resources in plant communities: a general theory of invasibility. *Journal of Ecology*, **88**, 528–534.

De Wreede, R.E. (1983) *Sargassum muticum* (Fucales, Phaeophyta): regrowth and interaction with *Rhodomela larix* (Ceramiales, Rhodophyta). *Phycologia*, **22** (2), 153–160.

Deysher, L. & Norton, T.A. (1982) Dispersal and colonization in *Sargassum muticum* (Yendo) Fensholt. *Journal of Experimental Marine Biology and Ecology*, **56** (2–3), 179–195.

Drake, J.M., Baggenstos, P. & Lodge, D.M. (2005) Propagule pressure and persistence in experimental populations. *Biology Letters*, **1**, 480–483.

Duggins, D.O. (1980) Kelp beds and sea otters: an experimental approach. *Ecology*, **61**, 447–453.

Duggins, D.O. (1983) Starfish predation and the creation of mosaic patterns in a kelp-dominated community. *Ecology*, **64**, 1610–1619.

Duggins, D.O., Eckman, J.E., Siddon, C.E. & Klinger, T. (2003) Population, morphometric and biomechanical studies of three understory kelps along a hydrodynamic gradient. *Marine Ecology Progress Series*, **265**, 57–76.

Foreman, R.E. (1977) Benthic community modification and recovery following intensive grazing by *Strongylocentrotus droebachiensis*. *Helgoländer wiss. Meeresunters*, **30**, 468–484.

Gentle, C.B. & Duggin, J.A. (1997) *Lantana camara* L. invasions in dry rainforest–open forest ecotones: the role of disturbances associated with fire and cattle grazing. *Australian Journal of Ecology*, **22**, 298–306.

Grevstad, F.S. (1999) Experimental invasions using biological control introductions: the influence of release size on the chance of population establishment. *Biological Invasions*, **1**, 313–323.

Griffith, B., Scott, J.M., Carpenter, J.W. & Reed, C. (1989) Translocation as a species conservation tool: status and strategy. *Science*, **245** (4917), 477–480.

Higgins, S.I. & Richardson, D.M. (1998) Pine invasions in the southern hemisphere: modelling interactions between organism, environment and disturbance. *Plant Ecology*, **135**, 79–93.

Hill, S.J., Tung, P.J. & Leishman, M.R. (2005) Relationships between anthropogenic disturbance, soil properties and plant invasion in endangered Cumberland plain woodland, Australia. *Austral Ecology*, **30**, 775–788.

Hobbs, R.J. & Huenneke, L.F. (1992) Disturbance, diversity, and invasion: implications for conservation. *Conservation Biology*, **6**, 324–337.

Kotanen, P.M. (1997) Effects of experimental soil disturbance on revegetation by natives and exotics on coastal Californian meadows. *Journal of Applied Ecology*, **34**, 631–644.

Larson, D.L. (2003) Native weeds and exotic plants: relationships to disturbance in mixed-grass prairie. *Plant Ecology*, **169**, 317–333.

Levin, P.S., Coyer, J.A., Petrik, R. & Good, T.P. (2002) Community-wide effects of nonindigenous species on temperate rocky reefs. *Ecology*, **83**, 3182–3193.

Levine, J.M. (2000) Species diversity and biological invasions: relating local process to community pattern. *Science*, **288**, 852–854.

Levine, J., Adler, P. & Yelenik, S. (2004) A meta-analysis of biotic resistance to exotic plant invasions. *Ecology Letters*, **7**, 975–989.

Levine, J.M. & D'Antonio, C.M. (1999) Elton revisited: a review of evidence linking diversity and invasibility. *Oikos*, **87**, 15–26.

Lockwood, J.L., Cassey, P. & Blackburn, T. (2005) The role of propagule pressure in explaining species invasions. *Trends in Ecology and Evolution*, **20**, 223–228.

Lonsdale, W.M. (1999) Global patterns of plant invasions and the concept of invasibility. *Ecology*, **80**, 1522–1536.

MacDonald, I., Loope, L., Usher, M. & Hamann, O. (1989) Wildlife conservation and the invasion of nature reserves by introduced species: a global perspective. *Biological Invasions: A global perspective* (eds J.A. Drake, H.A. Mooney, F. di Castri, R.H. Groves, F.J. Kruger, M. Rejmanek & M. Williamson), pp. 215–255. John Wiley & Sons, Chichester, UK.

Mack, M., Simberloff, D., Lonsdale, W., Evans, H., Clout, M. & Bazzaz, F. (2000) Biotic invasions: causes, epidemiology, global consequences, and control. *Ecological Applications*, **10**, 689–710.

Melbourne, B.A., Cornell, H.V., Davies, K.F., Dugaw, C.J., Elmendorf, S., Freestone, A.L., Hall, R.J., Harrison, S., Hastings, A., Holland, M., Holyoak, M., Lambrinos, J., Moore, K. & Yokomizo, H. (2007) Invasion in a heterogeneous world: resistance, coexistence or hostile takeover? *Ecology Letters*, **10**, 77–94.

Naeem, S., Knops, J., Tilman, D., Howe, K., Kennedy, T. & Gale, S. (2000) Plant diversity increases resistance to invasion in the absence of covarying extrinsic factors. *Oikos*, **91**, 97–108.

Nearshore Habitat Program (2001) *The Washington State ShoreZone Inventory*. Washington State Department of Natural Resources, Olympia, WA.

Norton, T.A. (1977) Ecological experiments with *Sargassum muticum*. *Journal of the Marine Biology Association of the United Kingdom*, **57**, 33–43.

Norton, T.A. & Deysher, L.E. (1988) The reproductive ecology of *Sargassum muticum* at different latitudes. *Reproduction, Genetics and Distributions of Marine Organisms: 23rd European Marine Biology Symposium, School of Biological Sciences, University of Wales, Swansea* (eds J.S. Ryland & P.A. Tyler), pp. 147–152. Olsen & Olsen, Fredensborg, Denmark.

Paine, R.T. & Vadas, R.L. (1969) The effects of grazing by sea urchins, *Strongylocentrotus* spp., on benthic algal populations. *Limnology and Oceanography*, **14**, 710–719.

Panetta, F.D. & Randall, R.P. (1994) An assessment of the colonizing ability of *Emex australis*. *Australian Journal of Ecology*, **19**, 76–82.

Parker, I.M. (2001) Safe site and seed limitation in *Cystisus scoparius* (Scotch Broom): invasibility, disturbance, and the role of cryptograms in a glacial outwash prairie. *Biological Invasions*, **3**, 323–332.

Pimentel, D., Zuniga, R. & Morrison, D. (2005) Update on the environmental and economic costs associated with alien-invasive species in the United States. *Ecological Economics*, **52**, 273–288.

Prieur-Richard, A. & Lavorel, S. (2000) Invasions: the perspective of diverse plant communities. *Austral Ecology*, **25**, 1–7.

Richardson, D.M. & Bond, W.J. (1991) Determinants of plant distribution: evidence from pine invasions. *American Naturalist*, **137**, 639–668.

Ruiz, G.M. & Carlton, J.T. (2003) Invasion vectors: a conceptual framework for management. *Invasive Species: Vectors and Management Strategies* (eds G.M. Ruiz & J.T. Carlton), pp. 459–504. Island Press.

Sanchez, I. & Fernandez, C. (2006) Resource availability and invasibility in an intertidal macroalgal assemblage. *Marine Ecology Progress Series*, **313**, 85–94.

Scagel, R.F. (1956) Introduction of a Japanese alga, *Sargassum muticum*, into the Northeast Pacific. *Washington Department of Fisheries, Fisheries Research Papers*, **1**, 1–10.

Simberloff, D. (1989) Which insect introductions succeed and which fail? *Biological Invasions: A global perspective, SCOPE 37* (eds J.A. Drake, H.A. Mooney, F. di Castri, R.H. Groves, F.J. Kruger, M. Rejmanek and M. Williamson), pp. 61–75. John Wiley & Sons, Chichester, UK.

Sjøtun, K., Eggereide, S.F. & Høisaeter, T. (2007) Grazer-controlled recruitment of the introduced *Sargassum muticum* (Phaeophyceae, Fucales) in northern Europe. *Marine Ecology Progress Series*, **342**, 127–138.

Stachowicz, J., Whitlatch, R. & Osman, R. (1999) Species diversity and invasion resistance in a marine ecosystem. *Science*, **286**, 1577–1579.

Thomsen, M.A., D'Antonio, C.M., Suttle, K.B. & Sousa, W.P. (2006) Ecological resistance, seed density and their interactions determine patterns of invasion in a California grassland. *Ecology Letters*, **9**, 160–170.

Travis, J.M.J., Hammershoj, M. & Stephenson, C. (2005) Adaptation and propagule pressure determine invasion dynamics: insights from a spatially
explicit model for sexually reproducing species. *Evolutionary Ecology
Research*, **7**, 37–51.

Vadas, R.L. (1968) *The ecology of* Agarum *and the kelp bed community.* PhD
Dissertation, University of Washington.

Valentine, J.P. & Johnson, C.R. (2003) Establishment of the introduced kelp
Undaria pinnatifida. Tasmania depends on disturbance to native algal
assemblages. *Journal of Experimental Marine Biology and Ecology*, **265**, 63–
90.

Valentine, J.P. & Johnson, C.R. (2005) Persistence of the exotic kelp *Undaria
pinnatifida* does not depend on sea urchin grazing. *Marine Ecology Progress
Series*, **285**, 43–55.

Veltman, C.J., Nee, S. & Crawley, M.J. (1996) Correlates of introduction
success in exotic New Zealand birds. *American Naturalist*, **147**, 542–557.

Von Holle, B. & Simberloff, D. (2005) Ecological resistance to biological
invasion overwhelmed by propagule pressure. *Ecology*, **86**, 3212–3218.

Williamson, M. (1989) Mathematical models of invasio. *Biological Invasions: A
global perspective* (ed. by J.A. Drake, H.A. Mooney, F. di Castri, R.H.
Groves, F.J. Kruger, M. Rejmanek and M. Williamson), pp. 329–350. John
Wiley & Sons, Chichester, UK.

Williamson, M. (1996) *Biological Invasions*. Chapman & Hall, London, UK.

Williamson, J. & Harrison, S. (2002) Biotic and abiotic limits to the spread of
exotic revegetation species. *Ecological Applications*, **12**, 40–51.

Received 12 June 2007; accepted 1 October 2007
Handling Editor: Jonathan Newman

Supplementary material

The following supplementary material is available for this
article:

Appendix S1. Detailed diet information for benthic, subtidal
mollusc species.

Appendix S2. Model parameter values and functions.

Appendix S3. Models for disturbance.

This material is available as part of the online article from:
http://www.blackwell-synergy.com/doi/abs/10.1111/j.1365-
2745.2007.01319.x
(This link will take you to the article abstract).

Please note: Blackwell Publishing is not responsible for the
content or functionality of any supplementary materials
supplied by the authors. Any queries (other than missing
material) should be directed to the corresponding author for
the article.

Answer pages

Task 2.1 Article headings and subheadings

Headings and subheadings for Kaiser et al. (2003)

Summary
Keywords
Introduction
Results
  Cloning of GmDmt1;1
  Gene expression
  Protein localisation
  Functional analysis in yeast
Discussion
  GmDmt1;1 can transport ferrous iron
  Specificity of GmDmt1;1
  Localisation and function of GmDmt1;1
  Regulation of GmDmt1;1 expression
  Conclusion
Experimental procedures
  Plant growth
  Isolation of GmDmt1;1
  Northern analysis
  Antibody generation and Western immunoblot analysis
  Symbiosome isolation and nodule membrane purification
  Functional expression in yeast
Acknowledgements
References

Headings and subheadings for Britton-Simmons and Abbott (2008)

Summary
Keywords
Introduction
Methods
  Study system
  The invader
  Field experiment
  Statistical analysis
  Model

Writing Scientific Research Articles: Strategy and Steps, 1st edition. By M. Cargill and P. O'Connor. Published 2009 by Blackwell Publishing, ISBN 978-1-4051-8619-3 (pb) and 978-1-4051-9335-1 (hb)

Results
Discussion
  Simulated urchin/mollusc disturbances
  Propagule pressure and invasion success
Conclusions
Acknowledgements
References
Supplementary material

Task 2.3 Structure of the PEAs

Kaiser et al. (2003) most closely resembles the AIRDaM diagram. Britton-Simmons and Abbott (2008) most closely resembles the AIMRaD diagram, with a separate Conclusions section added at the end.

Task 2.4 Prediction

… yielded a total of …	(R)
The aim of the work described …	(I)
… was used to calculate …	(M) or (R)
There have been few long-term studies of …	(I)
The vertical distribution of … was determined by …	(M) or (R)
This may be explained by …	(D)
Analysis was carried out using …	(M)
… was highly correlated with …	(R)

Task 3.1 Where would referees look?

See Table AP1.

Table AP1 Task 3.1: Where would referees look?

Referee criterion	Likely location of evidence
1. Is the contribution new?	I (also stated in A, but no room to demonstrate it there)
2. Is the contribution significant?	I and D (also stated in A)
3. Is it suitable for publication in the Journal?	T, I, A
4. Is the organization acceptable?	All
5. Do the methods and the treatment of results conform to acceptable scientific standards?	M and R
6. Are all conclusions firmly based in the data presented?	R compared to D and A
7. Is the length of the paper satisfactory?	All
8. Are all illustrations required?	Photographs
9. Are all the figures and tables necessary?	Figures and tables
10. Are figure legends and table titles adequate?	As above
11. Do the title and Abstract clearly indicate the content of the paper?	T, A and all
12. Are the references up to date, complete, and the journal titles correctly abbreviated?	Ref
13. Is the paper excellent, good, or poor?	All

Title A: Use of in situ ^{15}*N-labelling to estimate the total below-ground nitrogen of pasture legumes in intact soil-plant systems*

Information:

- The paper focuses on a particular method (*in situ* ^{15}N-labelling) and on results obtained using it.
- The parameter measured was total below-ground nitrogen.
- The measurement site/context was undisturbed systems involving both plants and soil.
- The plants used were pasture legumes.

Possible questions (many others are possible):

- Why is this method suitable to measure this parameter in this context?
- Did the method provide reliable measurements?
- How was the accuracy of the measurements verified?
- How many legumes were studied and how did the results vary between them?
- What soil types were involved?
- Could this method be used for other plant/soil systems?

Title B: Short- and long-term effects of disturbance and propagule pressure on a biological invasion

Information:

- The paper reports the effects of two factors (disturbance and propagule pressure) on one biological invasion.
- Results are reported over two time frames: short term and long term.
- The focus of the paper is on generalizations from the findings that apply to biological invasion in general (because no details are given in the title about the specific organisms or sites involved in this particular invasion).

Possible questions (many others are possible):

- What organisms and locations were involved in the invasion studied?
- What is the meaning of propagule pressure in this context?
- How are short term and long term defined in this paper?
- How do the specific results for this invasion provide evidence for the study of biological invasion in general?

Title C: The soybean NRAMP homologue, GmDMT1, is a symbiotic divalent metal transporter capable of ferrous iron transport

Information:

- The paper reports the function (ability to transport divalent metals) of a newly identified entity which is an NRAMP homologue found in soybeans.
- The work reported in the paper shows that the homologue can transport one particular type of iron (ferrous iron).
- The transport process is related to the symbiosis occurring in soybeans.

Possible questions (many others are possible):

- Why is the transport of ferrous iron significant in soybeans?
- How does the transport of divalent metals relate to the symbiosis?
- How was the function of this entity established?
- How does this finding contribute to the broader study of transporters?

Task 5.3 Identifying parts of figure legends

See Tables AP2 and AP3.

Table AP2 Task 5.3: Identifying parts of the legend from Britton-Simmons and Abbott (2008).

Sentence	Part
Number of *Sargassum muticum* (a) recruits and (b) adults in field experiment plots (900 cm^2).	Part 1
Propagule pressure is grams of reproductive tissue suspended over experimental plots at beginning of experiment.	Part 3
The average mass of an adult *S. muticum* (174 g) is indicated by an arrow.	Part 5
Data are means $\pm$ 1 SE ($n = 3$).	Part 4

Table AP3 Task 5.3: Identifying parts of the legend from Kaiser et al. (2003).

Sentence	Part
Uptake of Fe(II) by GmDmt1 in yeast.	Part 1
(a) Influx of ^{55}Fe^{2+} into yeast cells transformed with GmDmt1;1,	Part 1
fet3fet4 cells were transformed with GmDmt1;1-pFL61 or pFL61 and then incubated with 1 μM ^{55}FeCl$_3$ (pH 5.5) for 5- and 10-min periods.	Part 3
Data presented are means $\pm$ SE of ^{55}Fe uptake between 5 and 10 min from three separate experiments (each performed in triplicate).	Part 4
(b) Concentration dependence of ^{55}Fe influx into *fet3fet4* cells transformed with GmDmt1;1-pFL61 or pFL61.	Part 1
Data presented are means $\pm$ SE of ^{55}Fe uptake over 5 min (n=3).	Part 4
The curve was obtained by direct fit to the Michaelis-Menten equation.	Part 2
Estimated K_M and V_{MAX} for GmDmt1;1 were 6.4 $\pm$ 1.1 μM Fe(III) and 0.72 $\pm$ 0.08 nM Fe(III) min^{-1}mg^{-1} protein, respectively.	Part 2
(c) Effect of other divalent cations on uptake of ^{55}Fe^{2+} into *fet3fet4* cells transformed with pFL61-GmDMT1;1.	Part 1
Data presented are means $\pm$ SE of ^{55}Fe (10μM) uptake over 10 min in the presence and absence of 100μM unlabelled Fe^{2+}, Cu^{2+}, Zn^{2+} and Mn^{2+}.	Parts 3 and 4

Task 6.1 Separate location sentences in results sections

Kaiser et al. (2003): no separate location sentences occur.

Britton-Simmons and Abbott (2008): only one separate location sentence occurs, and it is written in the style of a highlight sentence:

We plotted the proportion of plots in each treatment combination that were successfully invaded as a function of propagule pressure (Fig. 3).

We can suggest two possible reasons for this choice:

1 this style allows the use of the active voice verb, in line with the more direct writing style preferred by these authors; and
2 the sentence is part of a longer section detailing what was done to answer the question posed at the start of the paragraph. The style of the sentence fits well with the way the other sentences have been constructed.

Task 7.1 Materials and methods organization

See Table AP4.

Table AP4 Task 7.1: Materials and methods organization.

Question	Britton-Simmons and Abbott (2008)	Kaiser et al. (2003)
1 What subheadings are used in the section?	Methods; Study system; The invader; Field experiment; Statistical analysis; Model	Experimental procedures; Plant growth; Isolation of GmDmt1;1; Northern analysis; Antibody generation and Western immunoblot analysis; Symbiosome isolation and nodule membrane purification
2.i How do the subheadings relate to the end of the Introduction?	Very clear relation to the last paragraph of the Introduction. Wordings related to each subheading have been used there in describing the principal activity of the study, and in almost the same order as the subheadings.	No specific relationship seen.
2.ii How do the subheadings relate to the subheadings in the Results section?	The last three subheadings come in the same order in which the Results are presented.	Results subheadings are not specifically related to Experimental procedure subheadings, but the order of the information in the Experimental procedure section follows closely the order in which the results are presented within that section.
3 Is the section easy for you to follow? Why? Or why not?	Aids to clarity include overview sentences at the start of paragraphs, before details are given.	Aids to clarity include frequent use of subheadings relating to order of information in Results, and use of purpose phrases to show why steps were taken in relation to the experimental aims.

Task 7.3 Active/passive sentences

See Table AP5.

Table AP5 Task 7.3: Active/passive sentences. Some examples of transformed sentences are given here. Compare them with the sentences you transformed.

PEA	Original sentence	Transformation
Kaiser et al. (2003)	Soybean seeds were inoculated at planting with *Bradyrhizobium japonicum* USDA 110 ... [passive] Subsequent PCR experiments identified a full-length 1849-bp cDNA ... [active]	We inoculated soybean seeds at planting with *Bradyrhizobium japonicum* USDA 110 ... [active] A full-length 1849-bp cDNA was identified in subsequent PCR experiments ... [passive]
Britton-Simmons and Abbott (2008)	Control plots were not altered in any way, ... [passive]	We did not alter control plots in any way, ... [active]
	Each holdfast produces as many as 18 laterals in the early spring, ... [active]	As many as 18 laterals are produced by each holdfast in early spring, ... [passive]

Task 7.4 Top-heavy passive sentences

Improved versions:

The soil water balance equation (Xin, 1986; Zhu and Niu, 1987) was used to compute actual evapotranspiration (T) for each crop, defined as the amount of precipitation for the period between sowing and harvesting the particular crop plus or minus the change in soil water storage in the 2m soil profile.

or

Actual evapotranspiration (T) for each crop was computed by the soil water balance equation (Xin, 1986; Zhu and Niu, 1987). This measure is defined as the amount of precipitation for the period between sowing and harvesting the particular crop plus or minus the change in soil water storage in the 2m soil profile.

Task 8.1 Introduction stages

See Tables AP6 and AP7.

Task 8.2 Introduction Stage 1 analysis

See Table AP8.

Table AP6 Task 8.1: Stages in the Introduction of Britton-Simmons and Abbott (2008).

Text	Stage
Biological invasions are a global problem with substantial economic (Pimentel *et al.* 2005) and ecological (Mack *et al.* 2000) costs. Research on invasions has provided important insights into the establishment, spread and impact of nonnative species. One key goal of invasion biology has been to identify the factors that determine whether an invasion will be successful (Williamson 1996). Accordingly, ecologists have identified several individual factors (e.g. disturbance and propagule pressure) that appear to exert strong controlling influences on the invasion process. However, understanding how these processes interact to regulate invasions remains a major challenge in ecology (D'Antonio *et al.* 2001; Lockwood *et al.* 2005; Von Holle & Simberloff 2005).	Stage 1 Stage 3 (broad research niche, claiming significance)
Propagule pressure is widely recognized as an important factor that influences invasion success (MacDonald *et al.* 1989; Simberloff 1989; Williamson 1996; Lonsdale 1999; Cassey *et al.* 2005). Previous studies suggest that the probability of a successful invasion increases with the number of propagules released (Panetta & Randall 1994; Williamson 1989; Grevstad 1999), with the number of introduction attempts (Veltman *et al.* 1996), with introduction rate (Drake *et al.* 2005), and with proximity to existing populations of invaders (Bossenbroek *et al.* 2001). Moreover, propagule pressure may influence invasion dynamics after establishment by affecting the capacity of non-native species to adapt to their new environment (Ahlroth *et al.* 2003; Travis *et al.* 2005). Despite its acknowledged importance, propagule pressure has rarely been manipulated experimentally and the interaction of propagule pressure with other processes that regulate invasion success is not well understood (D'Antonio *et al.* 2001; Lockwood *et al.* 2005).	Stage 2 Stage 3 (one component of the study, as indicated in the title)
Resource availability is a second key factor known to influence invasion success and processes that increase or decrease resource availability therefore have strong effects on invasions (Davis *et al.* 2000). Resource pre-emption by native species generates biotic resistance to invasion (Stachowicz *et al.* 1999; Naeem *et al.* 2000; Levine *et al.* 2004). Consequently, physical disturbance can facilitate invasions by reducing competition for limiting resources (Richardson & Bond 1991; Hobbs & Huenneke 1992; Kotanen 1997; Prieur-Richard & Lavorel 2000). In most communities disturbances occur via multiple mechanisms and the disturbances created by different agents vary in their intensity and frequency (D'Antonio *et al.* 1999). Recent empirical (Larson 2003; Hill *et al.* 2005) and theoretical (Higgins & Richardson 1998) studies suggest that not all types of disturbance have equivalent effects on the invasion process.	Stage 2

(Continued)

Table AP6 (*Continued*)

Text	Stage
Moreover, most of what we know about the effects of disturbance on invasions comes from short-term experimental studies. It is presently unclear how different disturbance agents influence long-term patterns of invasion.	Stage 3 (another component, as highlighted in the title)
In order for any invasion to be successful, propagule arrival must coincide with the availability of resources needed by the invading species (Davis *et al.* 2000). Therefore, the interaction between propagule pressure and processes that influence resource availability will ultimately determine invasion success (Brown & Peet 2003; Lockwood *et al.* 2005; Buckley *et al.* 2007).	Stage 2
In this study we used the invasion of shallow, subtidal kelp communities in Washington State by the Japanese seaweed *Sargassum muticum* as a study system to better understand the effects of propagule pressure and disturbance on invasion. In a factorial field experiment we manipulated both propagule pressure and disturbance in order to examine how these factors independently and interactively influence *S. muticum* establishment in the short term. We supplement the experimental results with a parameterized integrodifference equation model, which we use to examine how different natural disturbance agents influence the spread of *S. muticum* through the habitat in the longer term. Although a	Stage 4 (principal activities of the present study)
successful invasion clearly requires both establishment and spread of the invader, most studies have looked at just one of these processes (Melbourne *et al.* 2007). We take an integrative approach by employing both a short-term experiment and a longer-term model, allowing us to examine the effects of disturbance and propagule limitation on the entire invasion process.	Stage 5 (value of the present study, claiming significance)

Table AP7 Task 8.1: Stages in the Introduction of Kaiser et al. (2003).

Text	Stage
Legumes form symbiotic associations with N_2-fixing soil-borne bacteria of the *Rhizobium* family. The symbiosis begins when compatible bacteria invade legume root hairs, signalling the division of inner cortical root cells and the formation of a nodule. Invading bacteria migrate to the developing nodule by way of an 'infection thread', comprised of an invaginated cell wall. In the inner cortex, bacteria are released into the cell cytosol, enveloped in a modified plasma membrane (the peribacteroid membrane (PBM)), to form an organelle-like structure called the symbiosome, which consists of bacteroid, PBM and the intervening peribacteroid space (PBS; Whitehead and Day, 1997). The bacteria, subsequently, differentiate into the N_2-fixing bacteroid form. The symbiosis allows the access of legumes to atmospheric N_2, which is reduced to NH_4^+ by the bacteroid enzyme nitrogenase. In exchange for reduced N, the plant provides carbon to the nodules to support bacterial respiration, a low-oxygen environment in the nodule suitable for bacteroid nitrogenase activity, and all the essential nutritional elements necessary for bacteroid activity. Consequently, nutrient transport across the PBM is an important control mechanism in the promotion and regulation of the symbiosis.	Stage 1 (providing a context for the problem to be investigated)
	Stage 3 (broad research niche, claiming importance)
Micronutrients such as iron are essential for bacteroid activity and nodule development. The demand for iron increases during symbiosis (Tang et al., 1990), where the metal is used for the synthesis of various iron-containing proteins in both the plant and the bacteroids. In the plant fraction, iron is an important part of the heme moiety of leghemoglobin, which facilitates the diffusion of O_2 to the symbiosomes in the affected cell cytosol (Appleby, 1984). In bacteroids, there are many iron-containing proteins involved in N_2 fixation, including nitrogenase itself and cytochromes used in the bacteroid electron-transport chain. In the soil, iron is often poorly available to plants as it is usually in its oxidised form Fe(III), which is highly insoluble at neutral and basic pH. To compensate this, plants have developed two general strategies to gain access to iron from their localised environment. Strategy I involves initial reduction of Fe(III), while Strategy II involves secretion of phytosiderophores that aid in the solubilisation and uptake of Fe(III), followed by uptake of Fe(II) (Romheld, 1987). The mechanism(s) involved in bacteroid iron acquisition within the nodule have been investigated at the biochemical level, and three activities have been identified (Day et al., 2001). Fe(III) is transported across the PBM complexed with organic acids such as citrate, and accumulates in the PBS (Levier et al., 1996; Moreau et al., 1995), where it becomes bound to siderophore-like compounds (Wittenberg et al., 1996). Fe(III) chelate reductase activity has been measured on isolated PBM, and Fe(III) uptake into isolated symbiosomes is stimulated by Nicotinamide Adenine Dinucleotide (NADH), reduced form (Levier et al., 1996). However, Fe(II) is also readily transported across the PBM and has been found to be the favoured form of iron taken up by bacteroids (Moreau et al., 1998). The proteins involved in this transport have not yet been identified.	Stage 1 (another aspect of the context, iron, as indicated in the title)
	Stage 2 (aspects of the problem already investigated by others)
	Stage 3 (more specific research gap)

(Continued)

Table AP7 (*Continued*)

Text	Stage
Two classes of putative Fe(II)- transport proteins (Irt/Zip and Dmt/Nramp) have been identified in plants (Belouchi et al., 1997; Curie et al., 2000; Eide et al., 1996; Thomine et al., 2000). The Irt/Zip family was first identified in *Arabidopsis* by functional complementation of the yeast Fe(II) transport mutant DEY1453 (*fet3fet4*; Eide et al., 1996). *AtIrt1* expression is enhanced in roots when grown on low iron (Edie et al., 1996) and appears to be the main avenue for iron acquisition in *Arabidopsis* (Vert et al., 2002). Recently a soybean Irt/Zip isologue, GmZip1, was identified and localised to the PBM in nodules (Moreau et al., 2002). GmZip1 has been characterised as a symbiotic zinc transporter, which does not transport Fe(II).	Stage 2
The second class of iron-transport proteins consists of the Dmt/Nramp family of membrane transporters, which were first identified in mammals as a putative defense mechanism utilised by macrophages against mycobacterium infection (Supek et al., 1996; Vidal and Gros, 1994). Mutations in Nramp proteins in different organisms result invaried phenotypes including altered taste patterns in *Drosophila* (Rodrigues et al., 1995), microcyticanaemia (mk) in mice and Belgrade rats (Fleming et al., 1997) and loss of ethylene sensitivity in plants (Alonso et al., 1999). The rat and yeast NRAMP homologues (DCT1 and SMF1, respectively) have been expressed in *Xenopus* oocytes and shown to be broad-specificity metal ion transporters capable of Fe(II), among other divalent cations, transport (Chen et al., 1999; Gunshin et al., 1997). The plant homologue, AtNramp1, complements the growth defect of the yeast Fe(II) transport mutant DEY1453, while other *Arabidopsis* members do not (Curie et al., 2000; Thomine et al., 2000). Interestingly, AtNramp1 overexpression in *Arabidopsis* also confers tolerance to toxic concentrations of external Fe(II) (Curie et al., 2000), suggesting, perhaps, that it is localised intracellularly.	Stage 3 (building the gap) Stage 2
In this study we have identified a soybean homologue of the Nramp family of membrane proteins, GmDmt1;1. We show that GmDmt1;1 is a symbiotically enhanced plant protein, expressed in soybean nodules at the onset of nitrogen fixation, and is localised to the PBM. GmDmt1;1 is capable of Fe(II) transport when expressed in yeast. Together, the localisation and demonstrated activity of GmDmt1;1 in soybean nodules suggests that the protein is involved in Fe(II) transport and iron homeostasis in the nodule to support symbiotic N₂ fixation.	Stage 4 (principal activity of the study and conclusion reached)

Question	Kaiser et al. (2003)	Britton-Simmons and Abbott (2008)
Are some sentences written in the present tense? How many?	Yes, 8	Yes, 2
Are some sentences written in the present perfect tense? How many?	No	Yes, 3
Which tense is used more? Why do you think this is the case?	Present, because the focus of the content is explaining a biological process.	Present perfect, because the focus is on the developing field of research and the work others have done up to the present.
How many sentences contain references?	1 (of 8)	3 (of 5)
What kinds of sentences do not have references?	Sentences summarizing commonly accepted knowledge in the field.	Sentences that summarize the current state of knowledge in the field.

Task 8.3 Country to city in Stage 1

Kaiser et al. (2003)

What is the country? Legume symbiotic associations.
    The province? The peribacteroid membrane (PBM) and its role.
    The city? Nutrient transport across the PBM.

Britton-Simmons and Abbott (2008)

What is the country? Biological invasions.
    The province? Factors controlling the invasion process.
    The city? The interaction of the factors and processes.

Task 8.4 Identifying old or given information

Old information is underlined in the version below.

Legumes form symbiotic associations with N_2-fixing soil-borne bacteria of the *Rhizobium* family. The symbiosis begins when compatible bacteria invade legume root hairs, signalling the division of inner cortical root cells and the formation of a nodule. Invading bacteria migrate to the developing nodule by way of an 'infection thread', comprised of an invaginated cell wall. In the inner cortex, bacteria are released into the cell cytosol, enveloped in a modified plasma membrane (the peribacteroid membrane (PBM)), to form an organelle-like structure called the symbiosome, which consists of bacteroid, PBM and the intervening peribacteroid space (PBS; Whitehead and Day, 1997). The bacteria, subsequently, differentiate into the N_2-fixing bacteroid form. The symbiosis allows the access of legumes to atmospheric N_2, which is reduced to NH_4^+ by the bacteroid enzyme nitrogenase. In exchange for reduced N, the plant provides carbon to the nodules to support bacterial respiration, a low-oxygen environment in the nodule suitable for bacteroid nitrogenase activity, and all the essential nutritional elements necessary for bacteroid activity. Consequently, nutrient transport across the PBM is an important control mechanism in the promotion and regulation of the symbiosis.

Task 8.6 Identifying plagiarism

See Table AP9.

Table AP9 Task 8.6: Identifying plagiarism.

Plagiarized sentence in Version 2	Reason for the problem
However, this technique is not adaptable to all plants, particularly pasture species.	This sounds like the idea of the writer of the paragraph, but we know from Version 1 that it was originally the idea of Russell and Fillery (1996). Because there is no grammatical link between the two sentences, the reference in the first sentence does not apply to the second sentence. Note in Version 1 that the authors used both a grammatical link (they) and a tense marker (past tense *was not adaptable*) to indicate that the idea came from the cited work.

Task 8.7 Signal words for the research gap or niche

See Table AP10.

Table AP10 Task 8.7: Signal words for the research gap or niche.

McNeill et al. (1997)	Kaiser et al. (2003)	Britton-Simmons and Abbott (2008)
scarce, with little account taken of, is accordingly required, but, however	consequently, however, have not yet been identified, putative, appears to be	remains a major challenge, despite its acknowledged importance, rarely, is not well understood, it is presently unclear how, to better understand

Task 8.9 Stage 4 sentence templates

McNeill et al. (1997)

The experiments reported here were designed (i) to assess the use of [NP1] to [verb phrase], and (ii) to obtain quantitative estimates of [NP2].

In this study we have identified [NP1], [NP2]. We show that [NP2] is [NP3], expressed in [NP4] at the onset of [NP5], and is localised to [NP6].

Britton-Simmons and Abbott (2008)

In this study we used [NP1] as a study system to better understand the effects of [NP2] and [NP3] on [NP4]. In a [adjective] experiment we manipulated both [NP2] and [NP3] in order to examine how these factors [adverbs] influence [NP5] in the short term. We supplement the experimental results with [NP6], which we use to examine how different [NP7] influence [NP8] in the longer term.

Task 8.11 Topic sentence analysis

See Table AP11.

Table AP11 Task 8.11: Topic sentence analysis.

Reference	Topic sentence	Previous paragraph	Upcoming paragraph
Britton-Simmons and Abbott (2008)	Propagule pressure is widely recognized as an important factor that influences invasion success (references).	Refers to *propagule pressure* as one of two examples of factors influencing invasions.	Gives details of results of previous studies showing ways in which *propagule pressure* affects invasion success.
Kaiser et al. (2003)	Two classes of putative Fe(II)-transport proteins (Irt/Zip and Dmt/Nramp) have been identified in plants (references).	Ends by stating that the proteins involved are *unknown*, which links directly to *putative* (=possible candidates) in this sentence.	Gives details of research results on each of the two classes, in the same order in which they are referred to in the topic sentence (Irt/Zip and then Dmt/Nramp).

Task 8.12 Old information before new information

Here is the original text with inappropriately located new information underlined.

Pleuropneumonia (APP) surfaced in the Australian pig population during the first half of the 1980s and ten years later was regarded as one of the most costly and devastating diseases affecting the Australian pig industry. It can present as a dramatic clinical disease or as a chronic, production limiting disease in pig herds. <u>A sudden increase in the number of sick and coughing pigs and a sharp rise in mortalities among grower/finisher pigs</u> may herald an outbreak of APP in a herd. On the other hand, signs may be limited to a drop in growth rate and an increase in grade two pleurisy lesions in slaughter pigs.

Here is the revised wording of the problem sentence.

An outbreak of APP in a herd may be heralded by a sudden increase in the number of sick and coughing pigs and a sharp rise in mortalities among grower/finisher pigs.

N.B. Making this improvement involved changing an active voice verb (*may herald*) to a passive voice verb (*may be heralded*). This is one of the useful features of the active/passive verb system in English: it allows us to change the order of the information in a sentence. For science writers, this means we have an extra tool to enable us to get the given information at the start of each sentence.

Task 8.13 Revising top-heavy sentences

1 Original: In this project the *Rhizoctonia* populations of two field soils in the Adelaide Plains region of South Australia were characterised.

Suggested improvement options:

This project characterised the *Rhizoctonia* populations of two field soils in the Adelaide Plains region of South Australia.

or

The aim of this project was to characterise the *Rhizoctonia* populations of two field soils in the Adelaide Plains region of South Australia.

2 Original: A balance between deep and shallow rooting plants, heavy and light feeders, nitrogen fixers and consumers and an undisturbed phase is needed to achieve maximum benefit through rotation.

Suggested improvement options:

Maximum benefit through rotation can be achieved by using a balance between deep and shallow rooting plants, heavy and light feeders, nitrogen fixers and consumers and an undisturbed phase.

or

To achieve maximum benefit through rotation, it is necessary to have a balance between deep and shallow rooting plants, heavy and light feeders, nitrogen fixers and consumers and an undisturbed phase.

Task 9.2 Information elements in the Discussion section

See Tables AP12 and AP13.

Task 9.5 Negotiating strength of claims with verbs

See Table AP14. Notes on the suggested answers for Task 9.5 are given below.

- The relative strength of the meanings of these terms is to some extent a matter of opinion, and native speakers of English often disagree about the fine detail. The suggestions above represent the outcomes obtained by using this exercise in workshops over a number of years. We have included this exercise to raise your awareness of the issue of negotiating strength. We hope you will pay particular attention to how these words and forms are used in the papers you

Table AP12 Task 9.2, part 1: Information elements in the Discussion of Kaiser et al. (2003).

159

Sentences	Information element
The competition experiments shown in Figure 5(c) indicate that GmDmt1 can transport other divalent cations in addition to ferrous iron. Zinc, copper and manganese all inhibited iron uptake. The ability of GmDmt1;1 to enhance growth of the *zrt1zrt2* yeast mutant further suggests that the protein is not specific for iron transport.	2a. Restatement of one of the main findings, showing how it contributes to the main activity of the study
The preferred substrate *in vivo* may well depend on the relevant concentrations of divalent metals in the infected cell cytosol.	3. Speculation about the finding
This lack of specificity has been found with Nramp homologues from other organisms, including Nramp2 from mice. Despite this lack of specificity when expressed in heterologous systems, mutation of murine Nramp2 results in an anaemic phenotype, demonstrating that *in vivo* it is predominantly an iron transporter (Fleming *et al.*, 1997).	2b. Comparison with the findings of other researchers
Although GmDmt1;1 was able to complement the DEY1453 (*fet3fet4*) yeast mutant, the complementation was not robust and the growth media had to be supplemented with low concentrations of iron. Atlrt1, on the other hand, showed much better complementation and allowed growth of the mutant in the absence of added iron (Figure 4).	2a. Continued review of the finding
There are several possible reasons for the poorer growth with GmDmt1;1, including possible instability of GmDmt1;1 transcripts (perhaps because of the presence of the regulatory IRE element in the transcript).	3. Speculation about the findings

read in future, with the aim of fine-tuning your understanding of their usage in your own discipline area.

- *Can* indicates that the result has been recorded once, and is therefore possible. It does not make any claim for the likelihood that the result will be repeated.

- *Was stabilized* indicates that no claim is being made that the result is generalizable beyond the conditions of the experiment or study being reported. (In the sentence we are considering here, the verb in the next part of the sentence would have to be changed to the past also, if this alternative were used: *when free iron levels were low*).

- This usage of *should* indicates strong likelihood, and is often accompanied by a phrase giving the conditions under which the predicted event is likely to occur (as here with *when free iron levels are low*). It is not to be confused with the other usage of *should* to indicate a recommended action (e.g. You should wash your hands before meals.). The recommended action usage is much less common in scientific writing, although an example does occur in the Conclusions section of the PEA by Britton-Simmons and Abbott: "The model results demonstrate that caution *should be exercised* when extrapolating the results of short-term disturbance experiments over longer time intervals." This sentence demonstrates an important point: if an author wants to make a recommendation to the reader about future action, the reason for the recommendation needs to be very clear and well-supported. Here, it is the model results, already discussed in detail, that provide the support for making the *should* recommendation.

Table AP13 Task 9.2, part 2: Information elements in the Discussion of Britton-Simmons and Abbott (2008).

Sentences	Information elements
Our experimental results demonstrate that space- and propagule-limitation both regulate *S. muticum* recruitment. Our finding that *S. muticum* recruitment was positively related to propagule input is similar to those of two previous studies (Parker 2001; Thomsen *et al.* 2006), in which the propagule input of invasive plants was manipulated. In our control treatment space was limiting, a result that has also been found in previous studies of *S. muticum* recruitment (Deysher & Norton 1982; De Wreede 1983; Sanchez & Fernandez 2006). Consequently, increasing propagule pressure had a relatively weak effect on recruitment in undisturbed plots (Fig. 1a). However, when space limitation was alleviated by disturbing the plots, increasing propagule pressure caused a dramatic increase in recruitment (Fig. 1a). This suggests that in the presence of adequate substratum for settlement, propagule limitation becomes the primary factor controlling *S. muticum* recruitment. These results indicate that *S. muticum* recruitment under natural field conditions will be determined by the interaction between disturbance and propagule input.	2a. Restatement of the most important finding showing how it contributes to the main activity of the study 2b. Comparisons with the findings of other researchers 2a. Continued review of the important findings 5. Implications of the results (what they mean in the context of the broader field)

Table AP14 Task 9.5: Negotiating strength of claims with verbs, an exercise in ranking possible verb forms in a Discussion sentence in descending order of strength of claim.

					Weak
	implies		might be stabilized		
The	suggests	that	could be stabilized	by the	
presence	provides evidence	GmDmt1;1	may be stabilized	binding of IRPs	
of an IRE	indicates	mRNA	was stabilized	in soybean	↓
motif	shows		should be stabilized	nodules when free	
	demonstrates		is stabilized	iron levels	
				are low.	
					Strong

Task 10.1 Analyzing article titles

See Table AP15.

Task 11.1 Analyzing Summaries

See Tables AP16 and AP17.

Table AP15 Task 10.1: Analyzing article titles.

Question	Kaiser et al. (2003)	Britton-Simmons and Abbott (2008)
Is the title a noun phrase, a sentence, or a question?	Sentence	Noun phrase
How many words are used in the title?	16	13
What is the first idea in the title?	"The soybean NRAMP homologue, GmDMT1": the descriptor and name of the transporter discovered	"Short- and long-term effects"
Why do you think this idea has been placed first?	The descriptor comes first to show how this new discovery relates to what was previously known about the system under study.	This phrase highlights what is new and important about the work being reported.

Table AP16 Task 11.1: Summary analysis of Kaiser et al. (2003).

Summary sentences	Information elements
Iron is an important nutrient in N_2-fixing legume root nodules. Iron supplied to the nodule is used by the plant for the synthesis of leghemoglobin, while in the bacteroid fraction, it is used as an essential cofactor for the bacterial N_2-fixing enzyme, nitrogenase, and iron-containing proteins of the electron transport chain. The supply of iron to the bacteroids requires initial transport across the plant-derived peribacteroid membrane, which physically separated bacteroids from the infected plant cell cytosol.	Background
In this study we have identified *Glycine max divalent metal transporter 1 (GmDmt1)*, a soybean homologue of the NRAMP/Dmt1 family of divalent metal ion transporters.	Principal activity
GmDmt1 shows enhanced expression in soybean root nodules and is most highly expressed at the onset of nitrogen fixation in developing nodules.	Results*
Antibodies raised against a partial fragment of GmDmt1 confirmed its presence on the peribacteroid membrane (PBM) of soybean root nodules.	Method
GmDmt1 was able to both rescue growth and enhance ^{55}Fe(II) uptake in the ferrous iron transport deficient yeast strain (*fet3fet4*).	Results*
The results indicate that GmDmt1 is a nodule-enhanced transporter capable of ferrous iron transport across the PBM of soybean root nodules.	Conclusion
Its role in nodule iron homeostasis to support bacterial nitrogen fixation is discussed.	Another activity of the study/paper

*N.B. The first results sentence is written in the present tense: results obtained using the methods employed here are considered to represent outcomes seen as always being true. The third results sentence is written in the past tense; this indicates that the result represents an outcome specific to the experimental conditions used.

Table AP17 Task 11.1: Summary analysis of Britton-Simmons and Abbott (2008).

Summary sentences	Information elements
1. Invading species typically need to overcome multiple limiting factors simultaneously in order to become established, and understanding how such factors interact to regulate the invasion process remains a major challenge in ecology.	Background
2. We used the invasion of marine algal communities by the seaweed Sargassum muticum as a study system to experimentally investigate the independent and interactive effects of disturbance and propagule pressure in the short term. Based on our experimental results, we parameterized an integrodifference equation model, which we used to examine how disturbances created by different benthic herbivores influence the longer term invasion success of S. muticum.	Method + principal activity 1 Method + principal activity 2
3. Our experimental results demonstrate that in this system neither disturbance nor propagule input alone was sufficient to maximize invasion success. Rather, the interaction between these processes was critical for understanding how the S. muticum invasion is regulated in the short term.	Results
4. The model showed that both the size and spatial arrangement of herbivore disturbances had a major impact on how disturbance facilitated the invasion, by jointly determining how much space-limitation was alleviated and how readily disturbed areas could be reached by dispersing propagules.	Results
5. Synthesis. Both the short-term experiment and the long-term model show that S. muticum invasion success is co-regulated by disturbance and propagule pressure. Our results underscore the importance of considering interactive effects when making predictions about invasion success.	Results summary Conclusion/ recommendation

Task 13.1 The contributor's letter as sales pitch

See Figure AP1.

Task 17.1 Types of error

Part 2 Error type and likely effect on meaning

The suggested answers in Table AP18 are not absolute: they represent our judgement after considerable experience with EAL science text. However, it is important to remember two extra points:

- all these error types can on occasion affect the meaning of science writing to some extent: it can be useful to discuss specific examples to see how this happens; and

> Please find attached the manuscript "Arbuscular mycorrhizal associations of the southern Simpson Desert". This manuscript examines the mycorrhizal status of plants growing on the different soils of the dune-swale systems of the Simpson Desert. There have been few studies of the ecology of the plants in this desert and little is known about how mycorrhizal associations are distributed amongst the desert plants of Australia. We report the arbuscular mycorrhizal status of 47 plant species for the first time. The manuscript has been prepared according to the journal's Instructions for Authors. We believe that this new work is within the scope your journal and hope that you will consider this manuscript for publication in the *Australian Journal of Botany*.

Fig. AP1 Task 13.1 The contributor's letter as sales pitch. The highlighted words sell the novelty and significance of the manuscript to the editor.

Table AP18 Task 17.1, part 2: Suggested answers for assigning types of English language errors to three possible levels of effect on meaning.

Rarely/slightly affects meaning	Sometimes/moderately affects meaning	Often/seriously affects meaning
1	4	2
3	5	6
	7	8

Key to error types.

1 Incorrect usage of singular/plural forms (e.g. all tea leaves sample were oven dried).
2 Over-complex/inaccurate grammatical structures (e.g. This may be due to lower pH hinders dissolution of soil organic matter and decreases total dissolved Cu concentration because of Cu-organic complex reducing.).
3 Non-agreement of verbs and subjects (e.g. the results of this study suggests that …).
4 Incorrect choice of preposition (e.g. similar with the results of other researchers).
5 Non-standard usage of the articles *a/an* and *the* (e.g. the accumulation of Cu in human body).
6 Non-standard selection of modal verbs (e.g. *would* versus *will*, *can* versus *could* or *may*).
7 Incorrect choice of part of speech (e.g. drought resistance varieties).
8 Non-conventional selection of tense (e.g. present tense to refer to results of the study being reported).

- even the ones that affect meaning less can annoy readers, and lead to an (unwarranted) impression that the science is inaccurate because the English is inaccurate.

Part 3 Strategies for addressing different error types

Searches using the software ConcApp (see section 17.5) or a similar concordancing program can help you correct errors in categories 1, 4, and 7, and sometimes help with categories 2 and 6. Article errors (category 5) can be addressed using the flowchart (Figure 17.1) presented in section 17.6. The editing strategy using hard copy and a ruler, explained in section 15.2, item 7, is useful for finding errors in categories 1 and 3.

Task 17.2 Drafting a sentence template for Stage 4 of an Introduction

Britton-Simmons and Abbott (2008)

In this study we used [NP1] as a study system to better understand the effects of [NP2] and [NP3] on [NP4]. In a [adjectives] experiment we manipulated both [NP2] and [NP3] in order to examine how these factors [adverbs] influence [NP5] in [NP6]. We supplement the experimental results with [NP7], which we use to examine how [NP8] influence [NP9] in [NP10].

Kaiser et al. (2003)

In this study we have identified [NP1], [NP1a]. We show that [NP1a] is [NP2], expressed in [NP3] at [NP4], and is localised to [NP5]. [NP1a] is capable of [NP6] when expressed in [NP7].

Task 17.6 Generic noun phrases

Legumes form symbiotic associations with N_2-fixing soil-borne bacteria of the *Rhizobium* family. The symbiosis begins when compatible bacteria invade legume root hairs, signalling the division of inner cortical root cells and the formation of a nodule. Invading bacteria migrate to the developing nodule by way of an 'infection thread', comprised of an invaginated cell wall. In the inner cortex, bacteria are released into the cell cytosol, enveloped in a modified plasma membrane (the peribacteroid membrane (PBM)), to form an organelle-like structure called the symbiosome, which consists of bacteroid*, PBM* and the intervening peribacteroid space (PBS; Whitehead and Day, 1997). The bacteria, subsequently, differentiate into the N_2-fixing bacteroid form. The symbiosis allows the access of legumes to atmospheric N_2, which is reduced to NH_4^+ by the bacteroid enzyme nitrogenase. In exchange for reduced N, the plant provides carbon to the nodules to support bacterial respiration, a low-oxygen environment in the nodule suitable for bacteroid nitrogenase activity, and all the essential nutritional elements necessary for bacteroid activity. Consequently, nutrient transport across the PBM is an important control mechanism in the promotion and regulation of the symbiosis.

Task 17.7 Specific noun phrases

The specific noun phrases are shown with gray background.

Legumes form symbiotic associations with N_2-fixing soil-borne bacteria of the *Rhizobium* family. The symbiosis begins when compatible bacteria invade legume root hairs, signalling the division of inner cortical root cells and the formation of a nodule. Invading bacteria migrate to the developing nodule by way of an 'infection thread', comprised of an invaginated cell wall. In the inner cortex, bacteria are

*N.B. These two nouns are in fact specific here, but the two *the* articles have been omitted by the native-speaker authors. The use of articles is one of the most difficult areas of English grammar, and there is considerable debate about particular cases, even by so-called experts.

released into the cell cytosol, enveloped in a modified plasma membrane (the peribacteroid membrane (PBM)), to form an organelle-like structure called the symbiosome, which consists of bacteroid*, PBM* and the intervening peribacteroid space (PBS; Whitehead and Day, 1997). The bacteria, subsequently, differentiate into the N_2-fixing bacteroid form. The symbiosis allows the access of legumes to atmospheric N_2, which is reduced to NH_4^+ by the bacteroid enzyme nitrogenase. In exchange for reduced N, the plant provides carbon to the nodules to support bacterial respiration, a low-oxygen environment in the nodule suitable for bacteroid nitrogenase activity, and all the essential nutritional elements necessary for bacteroid activity. Consequently, nutrient transport across the PBM is an important control mechanism in the promotion and regulation of the symbiosis.

Task 17.8 Articles and plurals in a science paragraph

Propagule pressure is widely recognized as **an** important factor that influences invasion success. Previous studies suggest that **the** probability of successful invasion increases with **the** number of propagules released, with **the** number of introduction attempts, with introduction rate, and with proximity to existing populations of invaders. Moreover, propagule pressure may influence invasion dynamics after establishment by affecting **the** capacity of non-native species to adapt to their new environment. Despite its acknowledged importance, propagule pressure has rarely been manipulated experimentally and **the** interaction of propagule pressure with other processes that regulate invasion success is not well understood. (Britton-Simmons & Abbott 2008, p. 68)

N.B. The term *propagule pressure* remains generic throughout the paragraph – it refers to a concept, any or all instances of the concept, and the term *pressure* in this sense is uncountable – therefore no article is needed. *Introduction rate*, *proximity*, and *invasion success* are likewise generic and uncountable in this passage, so no article is needed.

Task 17.9 Punctuation with which and that

1 Lime, which raises the pH of the soil to a level more suitable for crops, is injected into the soil using a pneumatic injector.
2 *No additional punctuation required.*
3 Non-cereal phases, which are essential for the improvement of soil fertility, break disease cycles and replace important soil nutrients.
4 Senescence, which is the aging of plant parts, is caused by ethylene that the plant produces.
5 *No additional punctuation required.*
6 Seasonal cracking, which is a notable feature of this soil type, provides pathways at least 6 mm wide and 30 cm deep that assist in water movement into the subsoil.
7 *No additional punctuation required.*
8 Yellow lupin, which may tolerate waterlogging better than the narrow-leafed variety, has the potential to improve yields in this area.
9 *No additional punctuation required.*

*N.B. *the* omitted, as noted for Task 17.6.

References

Britton-Simmons, K.H. & Abbott, K.C. (2008) Short- and long-term effects of disturbance and propagule pressure on a biological invasion. *Journal of Ecology* 96, 68–77.

Flowerdew, J. & Li, Y. (2007) Language re-use among Chinese apprentice scientists writing for publication. *Applied Linguistics* 28, 440–65.

Kaiser, B.N., Moreau, S., Castelli, J., Thomson, R., Lambert, A., Bogliolo, S., Puppo, A., & Day, D.A. (2003) The soybean NRAMP homologue, GmDMT1, is a symbiotic divalent metal transporter capable of ferrous iron transport. *The Plant Journal* 35, 295–304.

Li, F., Zhao, S., & Geballe, G.T. (2000) Water use patterns and agronomic performance for some cropping systems with and without fallow crops in a semi-arid environment of northwest China. *Agriculture, Ecosystems and Environment* 79, 129–42.

McNeill, A.M., Zhu, C.Y., & Fillery, I.R.P. (1997) Use of *in situ* ^{15}N-labelling to estimate the total below-ground nitrogen of pasture legumes in intact soil-plant systems. *Australian Journal of Agricultural Research* 48, 295–304.

Sarpeleh, A., Wallwork, H., Catcheside, D.E.A., Tate, M.E., & Able, A.J. (2007) Protein-aceous metabolites from *Pyrenophora teres* contribute to symptom development of barley net blotch. *Phytopathology* 97, 907–15.

Weissberg, R. & Buker, S. (1990) *Writing Up Research: Experimental Research Report Writing for Students of English*. Prentice Hall Regents, Englewood Cliffs, NJ.

Index

173

Index